Solutions Manual to Accompany

Fundamentals of
Queueing Theory

Solutions Manual to Accompany

Fundamentals of Queueing Theory

Fourth Edition

Donald Gross

George Mason University
Fairfax, Virginia

John F. Shortle

George Mason University
Fairfax, Virginia

James M. Thompson

Freddie Mac Corporation
McLean, Virginia

Carl M. Harris

WILEY

A JOHN WILEY & SONS, INC., PUBLICATION

Published by John Wiley & Sons, Inc., Hoboken, New Jersey.
Published simultaneously in Canada.

For general information on our other products and services or for technical support, please contact our Customer Care Department within the United States at (800) 762-2974, outside the United States at (317) 572-3993 or fax (317) 572-4002.

Wiley also publishes its books in a variety of electronic formats. Some content that appears in print may not be available in electronic format. For information about Wiley products, visit our web site at www.wiley.com.

Library of Congress Cataloging-in-Publication Data:

Gross, Donald.

ISBN 978-0-470-07796-2

10 9 8 7 6 5 4 3 2

CONTENTS

CHAPTER 1

INTRODUCTION

1.3   The parameters are $\lambda = 40/$h and $1/\mu = 5.5\,$min. Using units of hours, $\mu = 60/5.5 \doteq 10.91/$h. The utilization should be less than 1, so $\lambda/c\mu \doteq 40/(10.91c)$, which implies that $c > 40/10.91 \doteq 3.67$. At least 4 are required to achieve steady state.

1.6   Let T be the total waiting time. If, when you arrive, the person in service is just about finished, then you wait on average eight service times (yours and the seven ahead of you) or $\mathrm{E}[T] = 8(2.5\,\text{min}) = 20\,\text{min}$. If, when you arrive, the person in service is just beginning, then you wait on average nine service times or $\mathrm{E}[T] = 9(2.5\,\text{min}) = 22.5\,\text{min}$. The average wait is somewhere in between.

Assuming the latter case, T is the sum of 9 IID normal random variables each with mean 2.5 and standard deviation 0.5. So T is a normal random variable with mean 22.5 and standard deviation $\sqrt{(9 \cdot 0.5^2)} = 1.5$. Then $\Pr\{T > 30\,\text{min}\} = \Pr\{Z > (30 - 22.5)/1.5\} = \Pr\{Z > 5\}$, where Z is a standard normal random variable. From standard normal tables, $\Pr\{Z > 5\} \doteq 0$.

Solutions Manual to Accompany Fundamentals of Queueing Theory, Fourth Edition.
By D. Gross, J. F. Shortle, J. M. Thompson, and C. M. Harris
Copyright © 2008 John Wiley & Sons, Inc.

1.9 Apply Little's law to the set of homes on the market. The average number of homes on the market is estimated as $L = 50$. The rate that homes enter the market is estimated as $\lambda = 5$ per week. By Little's law, a home is on the market for an average of $W = L/\lambda = 10$ weeks before it is sold. This assumes that the observed numbers are representative of long-term averages. Furthermore, it is assumed that you have no additional information that might change your estimate. For example, if you price your home at a very low price, you will probably sell it more quickly than the average.

1.12 The following table lists various statistics associated with each customer. "# in System" and "# in Queue" refer to the number of customers in the system and queue as seen by the arriving customer.

Customer # / Arrival Time	Service Start Time	Exit Time	Time in Queue	# in System	# in Queue
1	1.00	3.22	0.00	0	0
2	3.22	4.98	1.22	1	0
3	4.98	7.11	1.98	2	1
4	7.11	7.25	3.11	2	1
5	7.25	8.01	2.25	2	1
6	8.01	8.71	2.01	3	2
7	8.71	9.18	1.71	4	3
8	9.18	9.40	1.18	3	2
9	9.40	9.58	0.40	2	1
10	10.00	12.41	0.00	0	0
11	12.41	12.82	1.41	1	0
12	12.82	13.28	0.82	2	1
13	13.28	14.65	0.28	1	0
14	14.65	14.92	0.65	1	0
15	15.00	15.27	0.00	0	0

The values in the table are computed as follows:
- The exit time is the service-start time plus the service duration.
- The service-start time is the maximum of the exit time of the previous customer and the arrival time of the customer in question. (The first customer starts service immediately upon arrival.)
- The time in queue is the service-start time minus the arrival time.
- The number in system is the number of previously arriving customers whose exit time is after the arrival time of the customer in question.
- The number in queue is the number in system minus one, with a minimum value of zero.

$L_q^{(A)}$ is the average of the last column. $L_q^{(A)} = 12/15 = 0.8$. L_q is the total person minutes spent in the queue (the sum of the "Time in Queue" column) divided by the total time interval. $L_q = 17.02/15.27 = 1.1146$. Note that $L_q \neq L_q^{(A)}$.

1.15

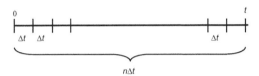

Divide the interval $[0, t]$ into n subintervals of length Δt, so that $t = n \, \Delta t$. The probability of one arrival in a subinterval is

$$p \equiv \Pr\{\text{one arrival in } \Delta t\} = \lambda \Delta t + o(\Delta t) \approx \frac{\lambda t}{n}.$$

The probability of more than one arrival in a subinterval is $o(\Delta t)$, which can be made arbitrarily small. Assuming that there can be at most one arrival in a subinterval and using the assumption of independence of nonoverlapping intervals, the total number of arrivals in $[0, t]$ is the sum of n Bernoulli trials. This follows a binomial distribution:

$$
\begin{aligned}
b(x; n, p) &= \binom{n}{x} p^x (1 - p)^{n-x}, \quad x = 0, 1, \ldots, n \\
&= \frac{n(n-1) \cdots (n - x + 1)}{x!} p^x (1 - p)^n (1 - p)^{-x} \\
&= \frac{1 \cdot (1 - 1/n) \cdots (1 - \frac{x-1}{n})}{x!} (np)^x \left(1 - \frac{\lambda t}{n}\right)^n \left(1 - \frac{\lambda t}{n}\right)^{-x}.
\end{aligned}
$$

So,

$$\lim_{n \to \infty} b(x; n, p) = \frac{1}{x!} (\lambda t)^x e^{-\lambda t}, \quad x = 0, 1, \ldots,$$

which is the Poisson distribution.

1.18   First, assume that n is even. Then,

$$
\begin{aligned}
p_n(t) &= \Pr\{N(t) = n\} \\
&= \Pr\{n \text{ singles}\} + \Pr\{(n - 2) \text{ singles and 1 double}\} \\
&\quad + \Pr\{(n - 4) \text{ singles and 2 doubles}\} + \cdots + \Pr\{n/2 \text{ doubles}\}.
\end{aligned}
$$

Then,

$$p_n(t) = \frac{e^{-\lambda t}(\lambda t)^n}{n!}p^n + \binom{n-1}{1}e^{-\lambda t}\frac{(\lambda t)^{n-1}}{(n-1)!}p^{n-2}(1-p)$$

$$+ \binom{n-2}{2}e^{-\lambda t}\frac{(\lambda t)^{n-2}}{(n-2)!}p^{n-4}(1-p)^2 + \cdots$$

$$+ \binom{n-n/2}{n/2}e^{-\lambda t}\frac{(\lambda t)^{n-n/2}}{(n-n/2)!}p^{n-2(n/2)}(1-p)^{n/2}.$$

So,

$$p_n(t) = e^{-\lambda t}\left\{ \frac{(\lambda t)^n}{n!}p^n + \frac{(\lambda t)^{n-1}}{1!(n-2)!}p^{n-2}(1-p) \right.$$

$$\left. + \frac{(\lambda t)^{n-2}}{2!(n-4)!}p^{n-4}(1-p)^2 + \cdots + \frac{(\lambda t)^{n/2}}{(n/2)!}(1-p)^{n/2} \right\}$$

$$= e^{-\lambda t}\sum_{k=0}^{n/2}\frac{(\lambda t)^{n-k}}{k!(n-2k)!}p^{n-2k}(1-p)^k.$$

Similarly, if n is odd,

$$p_n(t) = \Pr\{N(t) = n\}$$
$$= \Pr\{n \text{ singles}\} + \Pr\{(n-2) \text{ singles and 1 double}\}$$
$$+ \Pr\{(n-4) \text{ singles and 2 doubles}\}$$
$$+ \cdots + \Pr\{1 \text{ single and } (n-1)/2 \text{ doubles}\}.$$

Proceeding in the same manner gives

$$p_n(t) = e^{-\lambda t}\sum_{k=0}^{\lfloor n/2 \rfloor}\frac{(\lambda t)^{n-k}}{k!(n-2k)!}p^{n-2k}(1-p)^k.$$

1.21   Let

$$Q = \begin{pmatrix} -\lambda & \lambda & 0 & 0 & 0 & \cdots \\ 0 & -\lambda & \lambda & 0 & 0 & \cdots \\ 0 & 0 & -\lambda & \lambda & 0 & \cdots \\ \vdots & \vdots & \vdots & \vdots & \vdots & \end{pmatrix}.$$

Then $\mathbf{p}'(t) = \mathbf{p}(t)Q$ gives

$$(p_0'(t), p_1'(t), \ldots) = (p_0(t), p_1(t), \ldots)\begin{pmatrix} -\lambda & \lambda & 0 & 0 & 0 & \cdots \\ 0 & -\lambda & \lambda & 0 & 0 & \cdots \\ 0 & 0 & -\lambda & \lambda & 0 & \cdots \\ \vdots & \vdots & \vdots & \vdots & \vdots & \end{pmatrix},$$

which yields

$$p_0'(t) = -\lambda p_0(t),$$
$$p_1'(t) = \lambda p_0(t) - \lambda p_1(t),$$
$$p_2'(t) = \lambda p_1(t) - \lambda p_2(t),$$
$$\vdots$$
$$p_n'(t) = \lambda p_{n-1}(t) - \lambda p_n(t).$$

1.24

$$P(T_{(1)} \le t) = 1 - (1 - t)^n$$
$$P(nT_{(1)} \le t) = P\left(T_{(1)} \le \frac{t}{n}\right) = 1 - \left(1 - \frac{t}{n}\right)^n.$$

Since

$$\lim_{n \to \infty} \left(1 - \frac{t}{n}\right)^n = e^{-t},$$

then $\lim_{n \to \infty} P(nT_{(1)} \le t) = 1 - e^{-t}$, the exponential CDF.

CHAPTER 2

SIMPLE MARKOVIAN QUEUEING MODELS

2.3 Let N be the number of customers in the system in steady state. Let $p_n = \Pr\{N = n\}$. The generating function for $\{p_n\}$ was found in the text to be

$$P(z) = \sum_{n=0}^{\infty} p_n z^n = (1 - \rho)(1 - \rho z)^{-1}.$$

So,

$$P'(z) = \rho(1 - \rho)(1 - \rho z)^{-2},$$
$$P''(z) = 2\rho^2(1 - \rho)(1 - \rho z)^{-3},$$
$$P''(1) = 2\rho^2(1 - \rho)^{-2}.$$

Now,

$$P''(1) = \sum_{n=1}^{\infty} n(n - 1)p_n = \mathrm{E}[N^2] - \mathrm{E}[N].$$

Solutions Manual to Accompany Fundamentals of Queueing Theory, Fourth Edition.
By D. Gross, J. F. Shortle, J. M. Thompson, and C. M. Harris
Copyright © 2008 John Wiley & Sons, Inc.

So,

$$\mathrm{Var}[N] = \mathrm{E}[N^2] - (\mathrm{E}[N])^2 = P''(1) + \mathrm{E}[N] - (\mathrm{E}[N])^2$$

$$= \frac{2\rho^2}{(1-\rho)^2} + \frac{\rho}{1-\rho} - \frac{\rho^2}{(1-\rho)^2} = \frac{\rho^2}{(1-\rho)^2} + \frac{\rho - \rho^2}{(1-\rho)^2}$$

$$= \frac{\rho}{(1-\rho)^2}.$$

2.6

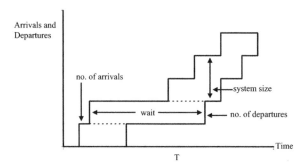

Adding up the horizontal blocks gives the total customer waiting minutes. Thus

$$W = \frac{\text{area under curve}}{\text{\# of customers in time } T}.$$

Adding up the vertical blocks gives the total system-size minutes. Thus

$$L = \frac{\text{area under curve}}{T} = \frac{W \cdot (\text{\# of customers in time } T)}{T}.$$

In steady state,

$$\frac{\text{\# of customers in time } T}{T} = \text{customer arrival rate} = \lambda.$$

This gives $L = \lambda W$.

For any finite time not coinciding with a busy cycle we have the error of neglecting (or including) customers still in the system. For our result to be valid for steady state we must show that this error $\to 0$ as $T \to \infty$. If all waits are finite this is obvious. But not so obvious for general waits: see Stidham, *Operations Research*, vol. 20, No. 6, Nov.-Dec. 1972.

2.9 Using QtsPlus

M/M/1: POISSON ARRIVALS TO A SINGLE EXPONENTIAL SERVER

Input Parameters:

Arrival rate (λ)	3.
Mean service time ($1/\mu$)	0.291667

Plot Parameters:

Maximum size for probability chart	10
Total time horizon for probability plotting	5.

Results:

Mean interarrival time ($1/\lambda$)	0.333333333
Service rate (μ)	3.428571429
Server utilization (ρ)	87.50%
Mean number of customers in the system (L)	7
Mean number of customers in the queue (Lq)	6.125
Expected non-empty queue size (Lq')	8
Mean waiting time (W)	2.333333333
Mean waiting time in the queue (Wq)	2.041666667
Mean length of busy period (B)	2.333333333

E[Daily Costs] $= 375 + 25L = 375 + 25(7) = 550$. Using QtsPlus again with the reduced service time, we have

M/M/1: POISSON ARRIVALS TO A SINGLE EXPONENTIAL SERVER

Input Parameters:

Arrival rate (λ)	3.
Mean service time ($1/\mu$)	0.25

Plot Parameters:

Maximum size for probability chart	10
Total time horizon for probability plotting	5.

Results:

Mean interarrival time ($1/\lambda$)	0.333333333
Service rate (μ)	4
Server utilization (ρ)	75.00%
Mean number of customers in the system (L)	3
Mean number of customers in the queue (Lq)	2.25
Expected non-empty queue size (Lq')	4
Mean waiting time (W)	1
Mean waiting time in the queue (Wq)	0.75
Mean length of busy period (B)	1

With the reduced service time, $L = 3$ and to find the new daily operating cost, say, K, we need to solve $K + 25(3) = 550$, so that $K = 550 - 75 = 475$.

2.12 Let $r = \lambda/\mu$ and $\rho = \lambda/\mu c$. Then $p_c = (r^c/c!)p_0$, so $p_0 = (c!/r^c)p_c$. From (2.31),

$$
p_n = \begin{cases} \dfrac{r^n}{n!} \dfrac{c!p_c}{r^c} = \dfrac{r^{n-c}c!\,p_c}{n!}, & \text{for } 1 \leq n \leq c, \\[4mm] \dfrac{r^n}{c^{n-c}c!} \dfrac{c!p_c}{r^c} = \rho^{n-c}p_c, & \text{for } n \geq c. \end{cases}
$$

From (2.33),

$$L_q = \left(\frac{r^c \rho}{c!(1 - \rho)^2} \right) p_0 = \left(\frac{\rho}{(1 - \rho)^2} \right) p_c.$$

2.15

(a) The utilization is $\rho = \lambda/\mu c = r/2 = 0.99$. So the offered load is $r = 2 \cdot 0.99 = 1.98$. With 3 servers the utilization is $r/3 = 0.66$. Therefore the percent idle time is 34%.

(b) The new service rate is $\mu^* = 0.8 \cdot \mu$. So the new utilization is

$$\rho^* = \frac{\lambda}{3\mu^*} = \frac{\lambda}{3 \cdot 0.8 \cdot \mu} = \frac{r}{2.4} = \frac{1.98}{2.4} = 0.825.$$

So the percent idle time is 17.5%.

(c) The new average service time for the two servers is

$$\frac{1}{\mu'} \equiv 0.8\frac{1}{\mu}.$$

So the new utilization is

$$\rho' = \frac{\lambda}{2\mu'} = 0.8\frac{\lambda}{2\mu} = 0.4r = 0.4 \cdot 1.98 = 0.792.$$

So the percent idle time is 20.8%.

2.18   Using the QtsPlus $M/M/c$ module with $c = 2$ we get the following:

M/M/c: POISSON ARRIVALS TO MULTIPLE EXPONENTIAL SERVERS

Input Parameters:

Arrival rate (λ)	60.
Mean service time ($1/\mu$)	0.022222
Number of servers in the system (c)	2

Plot Parameters:

Maximum size for probability chart	15
Total time horizon for probability plotting	2.

Results:

Mean interarrival time ($1/\lambda$)	0.016667
Service rate (μ)	45.
Average # arrivals in mean service time (r)	1.333333
Server utilization (ρ)	66.67%
Fraction of time all servers are idle (p_0)	0.2
Mean number of customers in the system (L)	2.4
Mean number of customers in the queue (Lq)	1.066667
Mean wait time (W)	0.04
Mean wait time in the queue (Wq)	0.017778

Customer Size Distribution				Waiting Time Distributions	
n	prob(n)	CDF(n)		t	Wq(t)
0	0.200000	0.200000		0.00	0.466667
1	0.266667	0.466667		0.01	0.604897
2	0.177778	0.644444		0.02	0.707300
3	0.118519	0.762963		0.03	0.783163
4	0.079012	0.841975		0.04	0.839363
5	0.052675	0.894650		0.05	0.880997

From the table, $\Pr(N \geq 2) = 1 - \Pr(N \leq 1) \doteq .5333$, and $\Pr(N \geq 4) = 1 - \Pr(N \leq 3) \doteq .2370$. Also, $\Pr\{T_q > .01\} \doteq 1 - .6049 = .3951$, and $\Pr\{T_q > .03\} \doteq 1 - .7832 = .2168$.

2.21 Using the QtsPlus $M/M/c$ module and using time units as weeks, we get that the expected time spent by an accident in and waiting for evaluation is about 3.4 weeks.

M/M/c: POISSON ARRIVALS TO MULTIPLE EXPONENTIAL SERVERS

Input Parameters:

Arrival rate (λ)	6.673077
Mean service time ($1/\mu$)	3.
Number of servers in the system (c)	23

Plot Parameters:

Maximum size for probability chart	15
Total time horizon for probability plotting	1.

Results:

Mean interarrival time ($1/\lambda$)	0.149856
Service rate (μ)	0.333333
Average # arrivals in mean service time (r)	20.019231
Server utilization (ρ)	87.04%
Fraction of time all servers are idle (p_0)	0.
Mean number of customers in the system (L)	22.82993
Mean number of customers in the queue (Lq)	2.810699
Mean wait time (W)	3.4212
Mean wait time in the queue (Wq)	0.4212

2.24 The first type of facility is an $M/M/2$ queue. Using QtsPlus, we get the following:

M/M/c: POISSON ARRIVALS TO MULTIPLE EXPONENTIAL SERVERS

Input Parameters:

Arrival rate (λ)	0.025
Mean service time ($1/\mu$)	30.
Number of servers in the system (c)	2

Plot Parameters:

Maximum size for probability chart	15
Total time horizon for probability plotting	2.

Results:

Mean interarrival time ($1/\lambda$)	40.
Service rate (μ)	0.033333
Average # arrivals in mean service time (r)	0.75
Server utilization (ρ)	37.50%
Fraction of time all servers are idle (p_0)	0.454545
Mean number of customers in the system (L)	0.872727
Mean number of customers in the queue (Lq)	0.122727
Mean wait time (W)	34.909091
Mean wait time in the queue (Wq)	4.909091

Thus (a) $L \doteq 0.873$ trucks and (b) $W \doteq 34.91$ min. Now the second type of facility is an $M/M/1$ queue. Using QtsPlus, we get the following:

M/M/1: POISSON ARRIVALS TO A SINGLE EXPONENTIAL SERVER

Input Parameters:

Arrival rate (λ)	0.025
Mean service time ($1/\mu$)	15.

Plot Parameters:

Maximum size for probability chart	20
Total time horizon for probability plotting	10.

Results:

Mean interarrival time ($1/\lambda$)	40
Service rate (μ)	0.066666667
Server utilization (ρ)	37.50%
Mean number of customers in the system (L)	0.6
Mean number of customers in the queue (Lq)	0.225
Expected non-empty queue size (Lq')	1.6
Mean waiting time (W)	24
Mean waiting time in the queue (Wq)	9
Mean length of busy period (B)	24

Thus (a) $L = 0.6$ trucks and (b) $W = 24$ min. For part (c), let C be the cost per minute of operating the single facility. Let L_1 and L_2 be the average number in the system for the single and dual facilities, respectively. In order for there to be no difference between the two systems regarding operating costs, we must have $C + 2L_1 = 1 + 2L_2$ (\$/min). Then, $C = 1 + 2(L_2 - L_1) \doteq 1 + 2(0.873 - 0.6) \doteq \$1.55$ per minute.

2.27 Let L_1 and L_2 be the average system sizes for the $M/M/1$ and $M/M/2$ queues, respectively. From Problem 2.26(a) we have $L_2 = 2\rho + 2\rho^3/(1 - \rho^2)$. We know that $L = L_q + r$ (Table 1.2), so that $L_2 = L_{q2} + 2\rho$, and thus $L_{q2} = L_2 - 2\rho = 2\rho^3/(1 - \rho^2)$. Now $L_{q1} = \rho^2/(1 - \rho)$, so that

$$\frac{L_{q1}}{L_{q2}} = \frac{1 + \rho}{2\rho} = \frac{1 + \rho}{\rho + \rho} > 1,$$

since $\rho < 1$. Hence

$$L_{q1} > L_{q2}.$$

But, from Problem 2.26(a), we saw that $L_2 > L_1$. Thus with one "super-server," customers spend more time in the queue, but less total time in the system. With two slower servers, customers spend less time in queue, but more total time in the system. Deciding which is preferable depends on the situation. In most people-related systems (e.g., fast-food establishments), it is queueing delay that is more annoying, so that the two slower servers would probably be preferred. If we are instead dealing with equipment to be repaired, it is total downtime that is most significant, so that the single superserver system would be preferred.

2.30   We use the QtsPlus $M/M/c/K$ module repeatedly to generate the following table:

Mean Queue Sizes for Rho = 1.5, 1.0, 0.8, 0.5

K	Lq(1.5)	Lq(1)	Lq(.8)	Lq(.5)
3	0	0	0	0
4	0.425	0.257	0.177	0.063
6	1.76	1.02	0.637	0.162
8	3.4	1.9	1.07	0.211
10	5.21	2.83	1.44	0.229
15	10.04	5.24	2.07	0.236
20	15	7.69	2.36	0.237
30	25	12.65	2.55	0.237
40	35	17.62	2.58	0.237
50	45	22.61	2.59	0.237
60	55	27.6	2.59	0.237

A graph of the table shows that for $\rho < 1$, as K gets large, L_q approaches the steady-state L_q for the $M/M/c$ queue, but for $\rho \geq 1$, L_q grows linearly.

Mean Queue Sizes for Rho = 1.5, 1.0, 0.8, 0.5

K	Lq(1.5)	Lq(1)	Lq(.8)	Lq(.5)
3	0	0	0	0
4	0.425	0.257	0.177	0.063
6	1.76	1.02	0.637	0.162
8	3.4	1.9	1.07	0.211
10	5.21	2.83	1.44	0.229
15	10.04	5.24	2.07	0.236
20	15	7.69	2.36	0.237
30	25	12.65	2.55	0.237
40	35	17.62	2.58	0.237
50	45	22.61	2.59	0.237
60	55	27.6	2.59	0.237

2.33   Using the $M/M/1/K$ QtsPlus module with $K = 9$, eight waiting seats plus the styling chair, we get the following:

M/M/1/K: POISSON ARRIVALS TO A SPACE-LIMITED SINGLE EXPONENTIAL SERVER

Input Parameters:

Arrival rate (λ)	5.
Mean service time ($1/\mu$)	.166667
Maximum capacity of system (K > 1)	9

Plot Parameters:

Total time horizon for probability plotting	20.
[ALL PROBABILITIES ARE PLOTTED!]	

Results:

Mean interarrival time ($1/\lambda$)	0.2
Effective arrival rate (λ_{eff})	4.80738622
Service rate (μ)	6
Traffic intensity (ρ) [NEED NOT BE < 1]	0.83333333
Server utilization (ρ_{eff}) [MUST BE < 100%]	80.12%
Fraction of time the server is idle (p_0)	0.19876896
Probability that the system is full (p_K)	0.03852276
Expected number turned away/unit time	0.19261378
Mean number of customers in the system (L)	3.07386216
Mean number of customers in the queue (Lq)	2.27263112
Mean waiting time (W)	0.63940404
Mean waiting time in the queue (Wq)	0.47273737

However, if there are only 4 waiting seats so that $K = 5$, we have the following:

M/M/1/K: POISSON ARRIVALS TO A SPACE-LIMITED SINGLE EXPONENTIAL SERVER

Input Parameters:

Arrival rate (λ)	5.
Mean service time ($1/\mu$)	.166667
Maximum capacity of system (K > 1)	5

Plot Parameters:

Total time horizon for probability plotting	20.
[ALL PROBABILITIES ARE PLOTTED!]	

Results:

Mean interarrival time ($1/\lambda$)	0.2
Effective arrival rate (λ_{eff})	4.49647127
Service rate (μ)	6
Traffic intensity (ρ) [NEED NOT BE < 1]	0.83333333
Server utilization (ρ_{eff}) [MUST BE < 100%]	74.94%
Fraction of time the server is idle (p_0)	0.25058812
Probability that the system is full (p_K)	0.10070575
Expected number turned away/unit time	0.50352873
Mean number of customers in the system (L)	1.97882762
Mean number of customers in the queue (Lq)	1.22941575
Mean waiting time (W)	0.44008457
Mean waiting time in the queue (Wq)	0.2734179

Thus for a K of 9, the effective arrival rate into the system is 4.807 customers per hour and for a K of 5, it is 4.496 customers per hour. So by adding the seat capacity, Cutt gains $4.807 - 4.496 = 0.311$ customers per hour for a daily profit increase of $0.311 \times 6 \times 6.75 = \$12.60$, which is far less than the $30 rent.

2.36   (a) Model the system as an $M/M/c/K$ queue with $\lambda = 15/\text{h}$, $\mu = 6/\text{h}$, $c = 3$, and $K = 24$. Using the QtsPlus software, it is found that $W_q \doteq 0.214$ hours or 12.9 minutes.

(b) Also using QtsPlus, it is found that $L \doteq 5.7$.

(c) The hourly cost is $L \cdot 60 \cdot \$0.03 + \lambda \cdot p_K \cdot \$20$, where p_K is the fraction of time the system is full. The first term gives the hourly cost of calls connected to your center. The second term gives the hourly cost of lost calls. Now, λ is fixed, but L and p_K both vary with K. Using QtsPlus, the hourly cost can be evaluated for various values of K (see below). The value of K that yields the lowest hourly cost is 39 (although higher values of K yield nearly the same hourly cost).

K	L	p_K	Hourly Cost
3	1.79	0.282	$87.8804
5	2.60	0.134	$45.7495
10	4.06	0.0390	$19.0226
15	4.96	0.0140	$13.1397
30	5.88	0.000856	$10.8460
35	5.95	0.000343	$10.8151
37	5.97	0.000238	$10.8124
38	5.97	0.000198	$10.8119
39	5.98	0.000165	$10.8118
40	5.98	0.000138	$10.8120
42	5.99	9.56×10^{-5}	$10.8128
50	6.01	2.22×10^{-5}	$10.8169
100	6.01	2.44×10^{-9}	$10.8202

2.39 For the $M/M/c$ queue,

$$p_0 = \left[\sum_{n=0}^{c-1} \frac{r^n}{n!} + \frac{r^c}{c!(1-\rho)} \right]^{-1}.$$

Then,

$$\lim_{c \to \infty} p_0 = e^{-r},$$

since $(r^c/c!) \to 0$ and $(1 - \rho) \to 1$ as $c \to \infty$. Also,

$$p_n = \begin{cases} \dfrac{r^n}{n!} p_0 & (0 \le n \le c) \\[2ex] \dfrac{r^n}{c^{n-c} c!} p_0 & (n \ge c) \end{cases}$$

Fixing n, as $c \to \infty$,

$$p_n \to \frac{r^n}{n!} p_0 = \frac{r^n}{n!} e^{-r}.$$

This is valid for each finite value of $n \ge 0$. These are the steady-state probabilities for the $M/M/\infty$ queue.

2.42   Using QtsPlus (Markov single-server, finite-source queue without spares), we can generate the following table for $M = 6, 7, \ldots$.

M	$M - 5$	$\Pr\{N \leq M - 5\}$
6	1	.81
7	2	.90
8	3	.94
9	4	.96

Thus we need to have 9 available to guarantee that 5 or more will be operating at least 95% of the time.

2.45   We have

$$q_n = \Pr\{n \text{ in system} \mid \text{arrival about to occur}\}$$
$$= \frac{\Pr\{n \text{ in system}\} \cdot \Pr\{\text{arrival about to occur} \mid n \text{ in system}\}}{\sum_n [\Pr\{n \text{ in system}\} \Pr\{\text{arrival about to occur} \mid n \text{ in system}\}]},$$

and

$$\lambda_n = \begin{cases} M\lambda & (0 \leq n < Y), \\ (M - n + Y) & (Y \leq n \leq Y + M), \\ 0 & (n > Y + M). \end{cases}$$

For $0 \leq n < Y$,

$$q_n = \lim_{\Delta t \to 0} \left\{ \frac{p_n[M\lambda\Delta t]}{\sum_{n=0}^{Y-1} p_n[M\lambda\Delta t] + \sum_{n=Y}^{Y+M} p_n[(M - n + Y)\lambda\Delta t]} \right\}$$
$$= \frac{Mp_n}{M - \sum_{n=Y}^{Y+M} (n - Y)p_n}.$$

Similarly, for $Y \leq n \leq Y + M - 1$,

$$q_n = \lim_{\Delta t \to 0} \left\{ \frac{p_n[(M - n + Y)\lambda\Delta t]}{\sum_{n=0}^{Y-1} p_n[M\lambda\Delta t] + \sum_{n=Y}^{Y+M} p_n[(M - n + Y)\lambda\Delta t]} \right\}$$
$$= \frac{(M - n + Y)p_n}{M - \sum_{n=Y}^{Y+M} (n - Y)p_n}.$$

To show $q_n(M) \neq p_n(M-1)$, we use a counterexample: Consider $M = 2$, $Y = 1$, $c = 1$, $\lambda/\mu = 1$. To find $q_n(M)$, we first find $p_n(M)$. Using (2.62), we get

$$p_0(M) = \tfrac{1}{11}, \quad p_1(M) = \tfrac{2}{11}, \quad p_2(M) = \tfrac{4}{11}, \quad p_3(M) = \tfrac{4}{11}.$$

Then using (2.65), which we have just derived,

$$q_0(M) = \tfrac{1}{5}, \quad q_1(M) = \tfrac{2}{5}, \quad q_2(M) = \tfrac{2}{5}.$$

To find $p_n(M-1)$, we again use (2.62) to get

$$p_0(M-1) = \tfrac{1}{3}, \quad p_1(M-1) = \tfrac{1}{3}, \quad p_2(M-1) = \tfrac{1}{3}.$$

Thus, $q_n(M) \neq p_n(M-1)$. To show for this particular example that $q_n(M) = p_n(Y-1)$, we find (using $M = 2$, $Y = 0$, $c = 1$, $\lambda/\mu = 1$) that

$$p_0(Y-1) = \tfrac{1}{5}, \quad p_1(Y-1) = \tfrac{2}{5}, \quad p_2(Y-1) = \tfrac{2}{5},$$

which is the same as $q_n(M)$.

2.48 We have M machines, Y spares, c technicians ($c \leq Y$). If $n = Y + 1$, the machines stop, so none can break. That is, the maximum number of failed machines is $n = Y + 1$. We have

$$\mu_n = \begin{cases} n\mu, & 0 \leq n \leq c, \\ c\mu, & n \geq c, \end{cases}$$

$$\lambda_n = \begin{cases} M\lambda, & 0 \leq n \leq Y, \\ 0, & n = Y + 1. \end{cases}$$

With $r = \lambda/\mu$,

$$p_n = \prod_{i=1}^{n} \frac{\lambda_{i-1}}{\mu_i} p_0,$$

so

$$p_n = \begin{cases} \dfrac{M^n}{n!} r^n p_0, & 0 \leq n < c, \\[2ex] \dfrac{M^n}{c^{n-c} c!} r^n p_0, & c \leq n \leq Y + 1, \end{cases}$$

where

$$p_0 = \left[\sum_{n=1}^{c-1} \frac{M^n}{n!} r^n + \sum_{n=c}^{Y+1} \frac{M^n}{c^{n-c} c!} r^n \right]^{-1}.$$

2.51 Assume that we use the lower speed when $n < k$ and the higher speed when $n \geq k$. Let C_3 denote the hourly cost of downtime of a lawn treater. Then,

$$\mathrm{E}[C] = \text{total costs/h} = C_1 \sum_{n=1}^{k-1} p_n + C_2 \left(1 - \sum_{n=0}^{k-1} p_n \right) + C_3 L.$$

Using (2.67), (2.68), and (2.69) with $\rho_1 = \frac{4}{3}$ and $\rho = \frac{2}{3}$,

$$p_0 = \left[\frac{(1-\rho_1)^k}{(1-\rho_1)} + \frac{\rho\rho_1^{k-1}}{(1-\rho)} \right]^{-1} = \left[\frac{1 - \left(\frac{4}{3}\right)^k}{-\frac{1}{3}} + \frac{\frac{2}{3}\left(\frac{4}{3}\right)^{k-1}}{\frac{1}{3}} \right]^{-1},$$

$$p_n = \rho_1^n p_0 = \left(\tfrac{4}{3}\right)^n p_0 \quad (0 \le n < k),$$

$$L = p_0 \left\{ \rho_1 \frac{[1 + (k-1)\rho_1^k - k\rho_1^{k-1}]}{(1-\rho_1)^2} + \rho\rho_1^{k-1} \frac{[k - (k-1)\rho]}{(1-\rho)^2} \right\}$$

$$= p_0 \left\{ \frac{\left(\frac{4}{3}\right)\left[1 + (k-1)\left(\frac{4}{3}\right)^k - k\left(\frac{4}{3}\right)^{k-1}\right]}{\left(-\frac{1}{3}\right)^2} \right.$$

$$\left. + \frac{\left(\frac{2}{3}\right)\left(\frac{4}{3}\right)^{k-1}\left[k - (k-1)\left(\frac{2}{3}\right)\right]}{\left(\frac{1}{3}\right)^2} \right\}.$$

When $k = 1$, we have

$$p_0 = \tfrac{1}{3}, \quad L = 2,$$
$$E[C(1)] = 110\left(1 - \tfrac{1}{3}\right) + 10 = \$83.33.$$

When $k = 2$, we have

$$p_0 = \tfrac{1}{5}, \quad L = 2.4, \quad p_1 = \tfrac{4}{15}, \quad \sum_{n=0}^{1} p_n = \tfrac{7}{15},$$
$$E[C(2)] = 25\left(\tfrac{4}{15}\right) + 110\left(1 - \tfrac{7}{15}\right) + 12 = \$77.33.$$

When $k = 3$, we have

$$p_0 = 0.13, \quad L = 2.96, \quad \sum_{n=1}^{2} p_n = 0.41, \quad \sum_{n=0}^{2} p_n = 0.54,$$
$$E[C(3)] = 25(0.41) + 110(1 - 0.54) + 14.8 = \$75.65$$

When $k = 4$, we have

$$p_0 = 0.09, \quad L = 3.6, \quad \sum_{n=1}^{3} p_n = 0.49, \quad \sum_{n=0}^{3} p_n = 0.58,$$
$$E[C(4)] = 25(0.49) + 110(1 - 0.58) + 18 = \$76.45.$$

The optimal value is $k = 3$.

2.54   Let the estimate of p_0 based on N terms be denoted by

$$p_0(N) = \left[\sum_{n=0}^{N} \frac{r^n}{(n!)^\alpha} \right]^{-1}.$$

When $\alpha \geq 1$,

$$\sum_{n=N+1}^{\infty} \frac{r^n}{(n!)^\alpha} \leq \sum_{n=N+1}^{\infty} \frac{r^n}{n!} = e^r - \sum_{n=0}^{N} \frac{r^n}{n!}. \tag{2.1}$$

We can find an N so that (2.1) is less than ϵ, since $\sum_{n=0}^{N} r^n/n!$ is converging to e^r. For such an N,

$$\epsilon + \sum_{n=0}^{N} \frac{r^n}{(n!)^\alpha} > \sum_{n=0}^{\infty} \frac{r^n}{(n!)^\alpha} = \frac{1}{p_0}$$

Hence,

$$\frac{1}{p_0(N)} = \sum_{n=0}^{N} \frac{r^n}{(n!)^\alpha} > \frac{1}{p_0} - \epsilon = \frac{1 - \epsilon p_0}{p_0}$$

and thus $p_0(N) < p_0/(1 - \varepsilon p_0) < p_0/(1 - \epsilon)$. Also, $p_0 < p_0(N)$ since

$$\left[\sum_{n=0}^{\infty} \frac{r^n}{(n!)^\alpha} \right]^{-1} < \left[\sum_{n=0}^{N} \frac{r^n}{(n!)^\alpha} \right]^{-1}.$$

In summary, $p_0 < p_0(N) < p_0/(1 - \epsilon) \approx p_0(1 + \epsilon)$.

2.57　The rate-balance diagram is below.

So,

$$p_n = \frac{(\lambda/n)(\lambda/(n-1)) \cdots (\lambda/2)(\lambda/1)}{\mu^n} p_0 = \frac{(\lambda/\mu)^n}{n!} p_0.$$

Normalizing,

$$1 = \sum_{n=0}^{\infty} p_n = p_0 \sum_{n=0}^{\infty} \frac{(\lambda/\mu)^n}{n!} = p_0 e^{\lambda/\mu}.$$

So,

$$p_n = e^{-(\lambda/\mu)} \frac{(\lambda/\mu)^n}{n!} = e^{-2} \frac{2^n}{n!}.$$

This is a Poisson distribution with mean 2. (An $M/G/\infty$ queue with $\lambda = 10$ and $\mu = 5$ also has the same steady-state distribution.)

2.60　From (2.77),

$$P(z,t) = \exp\left[\left(\frac{\lambda}{\mu}\right)(z-1)(1 - e^{-\mu t}) \right].$$

We also have

$$P(z,t) = \sum_{n=0}^{\infty} p_n(t)z^n$$

and

$$E[n(t)] = \left.\frac{\partial P(z,t)}{\partial z}\right|_{z=1} = \sum_{n=0}^{\infty} np_n(t).$$

So,

$$E[n(t)] = \left.\frac{\partial P(z,t)}{\partial z}\right|_{z=1}$$

$$= \left(\frac{\lambda}{\mu}\right)(1 - e^{-\mu t})\exp\left[\left(\frac{\lambda}{\mu}\right)(z-1)(1-e^{-\mu t})\right]\bigg|_{z=1}$$

$$= \left(\frac{\lambda}{\mu}\right)(1 - e^{-\mu t}).$$

2.63   **(a)**

$$\bar{f}(s) = \mathcal{L}\{f(t)\} = \int_0^{\infty} e^{-st}f(t)dt.$$

$$\mathcal{L}\{e^{-at}f(t)\} = \int_0^{\infty} e^{-st}e^{-at}f(t)dt = \int_0^{\infty} e^{-(s+a)t}f(t)dt = \bar{f}(s+a).$$

(b)

$$\mathcal{L}\left\{\sum_i a_i f_i(t)\right\} = \int_0^{\infty} e^{-st}\sum_i a_i f_i(t)dt = \sum_i a_i \int_0^{\infty} e^{-st}f_i(t)dt$$

$$= \sum_i a_i \bar{f}_i(s).$$

2.66   Let $X, X_2, \ldots, X_n$ be n independent random variables (not necessarily identically distributed) with moment generating functions

$$M_{X_i}(s), \qquad i = 1, 2, \ldots, n.$$

Let $Y = \sum_{i=1}^{n} X_i$. Then

$$M_Y(t) = E[e^{sY}] = E[e^{s(X_1+X_2+\cdots+X_n)}] = E[e^{sX_1}e^{sX_2}\cdots e^{sX_n}]$$

$$= E[e^{sX_1}]E[e^{sX_2}]E[e^{sX_3}]\cdots E[e^{sX_n}] = \prod_{i=1}^{n} M_{X_i}(s).$$

CHAPTER 3

ADVANCED MARKOVIAN QUEUEING MODELS

3.3   Using the QtsPlus bulk-input module, we get the following:

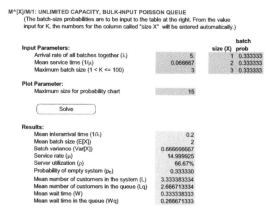

For the bulk-input model, $L_q \doteq 2.67$ and $L \doteq 3.33$. For the previous $M/M/1$ model, $L_q \doteq 1.33$ and $L \doteq 2$. The bulk-input model is more congested, even though the average arrival and service rates are the same.

3.6   Multiplying (3.10) by z^n and summing gives

$$\mu \sum_{n=0}^{\infty} p_{n+K} z^n - \lambda \sum_{n=0}^{\infty} p_n z^n - \mu \sum_{n=K}^{\infty} p_n z^n + \lambda \sum_{n=1}^{\infty} p_{n-1} z^n = 0.$$

Since

$$\sum_{n=0}^{\infty} p_{n+K} z^n = z^{-K} \sum_{n=0}^{\infty} p_{n+K} z^{n+K} = z^{-K} \left[P(z) - \sum_{n=0}^{K-1} p_n z^n \right],$$

this implies that

$$0 = \mu z^{-K} \left[P(z) - \sum_{n=0}^{K-1} p_n z^n \right] - \lambda P(z)$$

$$- \mu \left[P(z) - \sum_{n=0}^{K-1} p_n z^n \right] + \lambda z P(z)$$

$$= [\mu z^{-K} - (\lambda + \mu) + \lambda z] P(z) - \mu (z^{-K} - 1) \sum_{n=0}^{K-1} p_n z^n.$$

Thus

$$P(z) = \frac{(1 - z^K) \sum_{n=0}^{K-1} p_n z^n}{\rho z^{K+1} - (1 + \rho) z^K + 1}.$$

3.9   For state (n, i), we consider the following cases:

- $n \geq 2$ and $1 \leq i \leq k - 1$: The system leaves (n, i) when either an arrival occurs (with rate λ) or when a phase completion occurs (with rate $k\mu$). Possible ways to transition into this state are via an arrival from state $(n-1, i)$ (with rate λ) and via a phase completion from state $(n, i+1)$ (with rate $k\mu$). Equating rates gives

$$(\lambda + k\mu) p_{n,i} = \lambda p_{n-1,i} + k\mu p_{n,i+1}.$$

- $n \geq 2$ and $i = k$: The system leaves (n, k) when either an arrival occurs (with rate λ) or when a phase completion occurs (with rate $k\mu$). Possible ways to transition into this state are via an arrival from state $(n-1, k)$ (with rate λ) and via a phase completion from state $(n + 1, 1)$ (with rate $k\mu$). Equating rates gives

$$(\lambda + k\mu) p_{n,k} = \lambda p_{n-1,k} + k\mu p_{n+1,1}.$$

- $n = 1$ and $1 \leq i \leq k - 1$: The system leaves $(1, i)$ when either an arrival occurs (with rate λ) or when a phase completion occurs (with rate $k\mu$). The only way to transition into this state is via a phase completion from state $(1, i + 1)$ (with rate $k\mu$). Equating rates gives

$$(\lambda + k\mu) p_{1,i} = k\mu p_{1,i+1}.$$

○ $n = 1$ and $i = k$: The system leaves $(1, k)$ when either an arrival occurs (with rate λ) or when a phase completion occurs (with rate $k\mu$). Possible ways to transition into this state are via an arrival from state (0) (with rate λ) and via a phase completion from state $(2, 1)$ (with rate $k\mu$). Equating rates gives:

$$(\lambda + k\mu)p_{1,k} = \lambda p_0 + k\mu p_{2,1}.$$

○ $n = 0$: The system leaves (0) when an arrival occurs (with rate λ). The only way to transition into this state is via a phase completion from state $(1, 1)$ (with rate $k\mu$). Equating rates gives

$$\lambda p_0 = k\mu p_{1,1}.$$

3.12   The following table and plot show the various cases:

k	$\rho = .5$	$\rho = .7$	$\rho = .9$
1	.500	1.633	8.100
2	.375	1.225	6.075
3	.333	1.089	5.400
4	.313	1.021	5.063
5	.300	0.980	4.860
10	.275	0.898	4.455

Wq vs k for Various Rhos

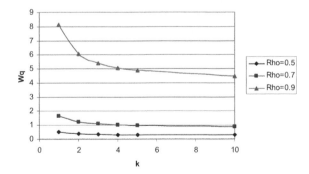

3.15   Let p_n denote the steady-state probability that there are n phases left in service, where $n = 0, 1, \ldots, k$. (This is like numbering the phases backwards, so a customer starts in phase k and ends in phase 1.) The rate balance equations are

$$\lambda p_0 = k\mu p_1,$$
$$k\mu p_n = k\mu p_{n+1} \quad (n = 1, 2, \ldots, k-1),$$
$$k\mu p_k = \lambda p_0.$$

So

$$p_1 = \frac{\lambda}{k\mu}p_0,$$

$$p_{n+1} = p_n \quad (n = 1, 2, \ldots, k-1),$$

$$p_k = \frac{\lambda}{k\mu}p_0.$$

This implies that

$$p_n = \frac{\lambda}{k\mu}p_0 \quad (n = 1, 2, \ldots, k).$$

The normalizing condition then gives

$$\left(1 + \sum_{i=1}^{k}\frac{\lambda}{k\mu}\right)p_0 = 1 \Rightarrow \left(1 + \frac{k\lambda}{k\mu}\right)p_0 = 1.$$

So, $p_0 = (1 + \rho)^{-1}$ and

$$p_n = \frac{\rho}{k(1 + \rho)} \quad (n = 1, 2, \ldots, k).$$

3.18 In an $M/E_k/1$ model, each customer requires k phases of service, where each phase is exponential with mean $1/k\mu$. Suppose when a customer arrives it is given k tickets and after completing each phase of service a ticket is taken. Before a customer can depart, it would have surrendered all k tickets. The number of phases of service in the system at any time would be the number of tickets outstanding. Now if we look at tickets as customers, then tickets arrive to the system in batches of k and each ticket is served according to an exponential service time with a mean of $1/k\mu$. Thus the model describing the ticket is $M^{[k]}/M/1$ with a mean service rate of $k\mu$.

3.21 (a) Let $\{A, n\}$ denote the system state where the teller is either serving a customer or available to serve customers and there are n customers in the system. Let $\{N, n\}$ denote the system state where the teller is *not* available (i.e., the teller is counting money) and there are n customers in the system. The diagram shows the transition rates between states.

The rate balance equations, for $n \geq 1$, are

$$(\lambda + \mu)p_{A,n} = \lambda p_{A,n-1} + \mu p_{A,n+1} + \gamma p_{N,n},$$
$$(\lambda + \gamma)p_{N,n} = \lambda p_{N,n-1}.$$

The boundary equations ($n = 0$) are

$$\lambda p_{A,0} = \gamma p_{N,0},$$
$$(\lambda + \gamma)p_{N,0} = \mu p_{A,1}.$$

(b)

$$L = \sum_{n=1}^{\infty} n(p_{A,n} + p_{N,n}), \qquad L_q = \sum_{n=1}^{\infty} (n-1)p_{A,n} + np_{N,n}.$$

3.24   From the data, the sample mean is 3.2 min and the sample variance is 2.5 min^2 (so the sample standard deviation is 1.581 min). Thus the sample CV (std.dev/mean) is $1.581/3.2 \doteq 0.494$. Since the CV is less than 1, we consider an Erlang. To find k, we know the Erlang-k CV is $1/\sqrt{k}$ and equating this to 0.494 yields a k of 4.098. Rounding to nearest integer we use $k = 4$. Using the M/E$_k$/1 module of QtsPlus we get the following:

M/E(k)/1: POISSON ARRIVALS TO A SINGLE ERLANG SERVER
Enter input and plot parameters, then press "Solve" button.

Input Parameters:

Arrival rate (λ)	0.3
Mean service time ($1/\mu$)	3.2
Erlang shape parameter (k)	4

Plot Parameters:

Maximum size for probability chart	10
Maximum time for Wq(t) plot	5

Results:

Mean interarrival time ($1/\lambda$)	3.333333
Service rate (μ)	0.3125
Server utilization (ρ)	96.00%
Probability for an empty system (p_0)	0.040000
Mean number of customers in the system (L)	15.36
Mean number of customers in the queue (Lq)	14.4
Mean waiting time (W)	51.2
Mean waiting time in the queue (Wq)	48.0
Mean length of busy period (B)	80.0

The average wait in queue is 48 min, and the average number in queue is 14.4.

3.27   The balance equations can be obtained by enumerating the possible state transitions and their associated probabilities (see table). For example, from the first entry in the table, the probability that the system is in state 0 at time $(t + \Delta t)$ is approximately

$$(1 - \lambda \Delta t)p_0 + (\mu \Delta t)p_{101} + (\mu \Delta t)p_{012}.$$

But this probability is also p_0. Equating these and dividing by Δt gives

$$\lambda p_0 = \mu p_{101} + \mu p_{012}.$$

System State		Event	Transition Prob.
At $t + \Delta t$	At t		
(0)	(0)	nothing	$(1 - \lambda\Delta t)p_0$
	$(1,0,1)$	class-1 departs	$(\mu\Delta t)p_{101}$
	$(0,1,2)$	class-2 departs	$(\mu\Delta t)p_{012}$
$(1,0,1)$	$(1,0,1)$	nothing	$[1 - (\lambda + \mu)\Delta t]p_{101}$
	(0)	class-1 arrives	$(\lambda_1\Delta t)p_0$
	$(2,0,1)$	class-1 departs	$(\mu\Delta t)p_{201}$
	$(1,1,2)$	class-2 departs	$(\mu\Delta t)p_{112}$
$(0,1,2)$	$(0,1,2)$	nothing	$[1 - (\lambda + \mu)\Delta t]p_{012}$
	(0)	class-2 arrives	$(\lambda_2\Delta t)p_0$
	$(1,1,1)$	class-1 departs	$(\mu\Delta t)p_{111}$
	$(0,2,2)$	class-2 departs	$(\mu\Delta t)p_{022}$
$(m,0,1)$	$(m,0,1)$	nothing	$[1 - (\lambda + \mu)\Delta t]p_{m,0,1}$
$m \geq 2$	$(m-1,0,1)$	class-1 arrives	$(\lambda_1\Delta t)p_{m-1,0,1}$
	$(m+1,0,1)$	class-1 departs	$(\mu\Delta t)p_{m+1,0,1}$
	$(m,1,2)$	class-2 departs	$(\mu\Delta t)p_{m,1,2}$
$(0,n,2)$	$(0,n,2)$	nothing	$[1 - (\lambda + \mu)\Delta t]p_{0,n,2}$
$n \geq 2$	$(0,n-1,2)$	class-2 arrives	$(\lambda_2\Delta t)p_{0,n-1,2}$
	$(1,n,1)$	class-1 departs	$(\mu\Delta t)p_{1,n,1}$
	$(0,n+1,2)$	class-2 departs	$(\mu\Delta t)p_{0,n+1,2}$
$(1,n,1)$	$(1,n,1)$	nothing	$[1 - (\lambda + \mu)\Delta t]p_{1,n,1}$
$n \geq 1$	$(1,n-1,1)$	class-2 arrives	$(\lambda_2\Delta t)p_{1,n-1,1}$
	$(2,n,1)$	class-1 departs	$(\mu\Delta t)p_{2,n,1}$
	$(1,n+1,2)$	class-2 departs	$(\mu\Delta t)p_{1,n+1,2}$
$(m,1,2)$	$(m,1,2)$	nothing	$[1 - (\lambda + \mu)\Delta t]p_{m,1,2}$
$m \geq 1$	$(m-1,1,2)$	class-1 arrives	$(\lambda_1\Delta t)p_{m-1,1,2}$
$(m,n,1)$	$(m,n,1)$	nothing	$[1 - (\lambda + \mu)\Delta t]p_{m,n,1}$
$m \geq 2$	$(m-1,n,1)$	class-1 arrives	$(\lambda_1\Delta t)p_{m-1,n,1}$
$n \geq 1$	$(m,n-1,1)$	class-2 arrives	$(\lambda_2\Delta t)p_{m,n-1,1}$
	$(m+1,n,1)$	class-1 departs	$(\mu\Delta t)p_{m+1,n,1}$
	$(m,n+1,2)$	class-2 departs	$(\mu\Delta t)p_{m,n+1,2}$
$(m,n,2)$	$(m,n,2)$	nothing	$[1 - (\lambda + \mu)\Delta t]p_{m,n,2}$
$m \geq 1$	$(m-1,n,2)$	class-1 arrives	$(\lambda_1\Delta t)p_{m-1,n,2}$
$n \geq 2$	$(m,n-1,2)$	class-2 arrives	$(\lambda_2\Delta t)p_{m,n-1,2}$

3.30 Without loss of generality, assume that $\mu_1 \leq \mu_2$, so that $\mu = \mu_1$ for the single-rate priority queue. From (3.36)

$$
\begin{aligned}
L_q^{(1)} &= \frac{\lambda_1 \rho}{\mu - \lambda_1} = \frac{\lambda_1(\lambda_1/\mu_1 + \lambda_2/\mu_1)}{\mu_1 - \lambda_1} = \frac{(\lambda_1/\mu_1)(\lambda_1/\mu_1 + \lambda_2/\mu_1)}{1 - \rho_1} \\
&= \frac{\lambda_1(\rho_1/\mu_1 + \lambda_2/\mu_1^2)}{1 - \rho_1} \geq \frac{\lambda_1(\rho_1/\mu_1 + \lambda_2/\mu_2^2)}{1 - \rho_1} \\
&= \frac{\lambda_1(\rho_1/\mu_1 + \rho_2/\mu_2)}{1 - \rho_1} = L_q^{(1)} \quad \text{for the unequal-rate case.}
\end{aligned}
$$

Similarly, starting from (3.36) with $\mu = \mu_1$

$$
\begin{aligned}
L_q^{(2)} &= \frac{\lambda_2 \rho}{(\mu - \lambda_1)(1 - \rho)} = \frac{\lambda_2(\lambda_1/\mu_1 + \lambda_2/\mu_1)}{(\mu_1 - \lambda_1)(1 - \lambda_1/\mu_1 - \lambda_2/\mu_1)} \\
&= \frac{(\lambda_2/\mu_1)(\lambda_1/\mu_1 + \lambda_2/\mu_1)}{(1 - \rho_1)(1 - \lambda_1/\mu_1 - \lambda_2/\mu_1)} = \frac{\lambda_2(\rho_1/\mu_1 + \lambda_2/\mu_1^2)}{(1 - \rho_1)(1 - \lambda_1/\mu_1 - \lambda_2/\mu_1)} \\
&\geq \frac{\lambda_2(\rho_1/\mu_1 + \lambda_2/\mu_2^2)}{(1 - \rho_1)(1 - \lambda_1/\mu_1 - \lambda_2/\mu_2)} = \frac{\lambda_2(\rho_1/\mu_1 + \rho_2/\mu_2)}{(1 - \rho_1)(1 - \rho)} \\
&= L_q^{(2)} \quad \text{for the unequal-rate case.}
\end{aligned}
$$

3.33 When $\mu_i \equiv \mu$, (3.43) becomes

$$
\begin{aligned}
W_q^{(i)} &= \frac{\sum_{k=1}^{r} \lambda_k/\mu^2}{\left(1 - \sum_{k=1}^{i-1} \lambda_k/\mu\right)\left(1 - \sum_{k=1}^{i} \lambda_k/\mu\right)} \\
&= \frac{\lambda}{\left(\mu - \sum_{k=1}^{i-1} \lambda_k\right)\left(\mu - \sum_{k=1}^{i} \lambda_k\right)}.
\end{aligned}
$$

This implies that

$$
\overline{W}_q \equiv \sum_{i=1}^{r} \frac{\lambda_i W_q^{(i)}}{\lambda} = \sum_{i=1}^{r} \frac{\lambda_i}{\left(\mu - \sum_{k=1}^{i-1} \lambda_k\right)\left(\mu - \sum_{k=1}^{i} \lambda_k\right)}.
$$

The sum of the first two terms ($i = 1$ and $i = 2$) is

$$
\begin{aligned}
&\frac{\lambda_1}{\mu(\mu - \lambda_1)} + \frac{\lambda_2}{(\mu - \lambda_1)(\mu - \lambda_1 - \lambda_2)} \\
&= \frac{1}{\mu - \lambda_1}\left[\frac{\lambda_1(\mu - \lambda_1 - \lambda_2) + \lambda_2(\mu - \lambda_1 + \lambda_1)}{\mu(\mu - \lambda_1 - \lambda_2)}\right] \\
&= \frac{\lambda_1 + \lambda_2}{\mu(\mu - \lambda_1 - \lambda_2)}.
\end{aligned}
$$

In a similar manner, adding the third term ($i = 3$) to this previous sum gives

$$
\frac{\lambda_1 + \lambda_2 + \lambda_3}{\mu(\mu - \lambda_1 - \lambda_2 - \lambda_3)}.
$$

Continuing in the same way eventually gives

$$\bar{W}_q = \frac{\sum_{k=1}^{r} \lambda_k}{\mu \left(\mu - \sum_{k=1}^{r} \lambda_k \right)} = \frac{\lambda}{\mu(\mu - \lambda)},$$

which is the result for the $M/M/1$ queue.

3.36   Using the QtsPlus software, the average time to process a general patient is $W = 1.1$ h, the same as the case with no preemption. The average time to process a class-1 patient decreases, the average time to process a class-2 patient decreases, and the average time to process a class-3 patient increases.

> **SINGLE-SERVER, MARKOV PREEMPTIVE MODEL WITH UP TO FIVE PRIORITY CLASSES**
> One service rate and preemption, with lower number having higher priority.
> Leave unneeded class arrival rates blank.
>
> **Input Parameters:**
>
> | | |
> |---|---|
> | Mean arrival rate for class 1 ($\lambda 1$) | 2. |
> | Mean arrival rate for class 2 ($\lambda 2$) | 3. |
> | Mean arrival rate for class 3 ($\lambda 3$) | 5. |
> | Mean arrival rate for class 4 ($\lambda 4$) | |
> | Mean arrival rate for class 5 ($\lambda 5$) | |
> | Overall mean service time ($1/\mu$) | 0.091667 |
>
> **Results:**
>
> | | |
> |---|---|
> | Overall arrival rate (λ) | 10 |
> | Mean interarrival time ($1/\lambda$) | 0.1 |
> | Overall service rate (μ) | 10.90909091 |
> | Server utilization (ρ) | 91.67% |
> | Probability of an empty system (p_0) | 0.083333 |
> | Mean number of customers in the system (L) | 10.55044479 |
> | Mean number of customers in the queue (Lq) | 9.633778127 |
> | Mean wait time (W) | 1.055044479 |
> | Mean wait time in the queue (Wq) | 0.963377813 |
> | **Priority Class #1:** | |
> | Expected time in the system (W1) | 0.112244898 |
> | Expected waiting time in the queue (Wq1) | 0.020578231 |
> | Expected number in the system (L1) | 0.224489796 |
> | Expected number in the queue (Lq1) | 0.041156463 |
> | **Priority Class #2:** | |
> | Expected time in the system (W2) | 0.186643119 |
> | Expected waiting time in the queue (Wq2) | 0.094976452 |
> | Expected number in the system (L2) | 0.559929356 |
> | Expected number in the queue (Lq2) | 0.284929356 |
> | **Priority Class #3:** | |
> | Expected time in the system (W3) | 1.953205128 |
> | Expected waiting time in the queue (Wq3) | 1.861538462 |
> | Expected number in the system (L3) | 9.766025641 |
> | Expected number in the queue (Lq3) | 9.307692308 |

3.39   Substituting $P_1(z) = \rho P_0(z) + (z\gamma/\mu)P_0'(z)$ into (3.51) gives

$$(\lambda \rho + \lambda)P_0(z) + (z\rho\gamma + z\gamma)P_0'(z) = \lambda P_0(z) + \gamma P_0'(z) + \lambda z\rho P_0(z)$$
$$+ \rho z^2 \gamma P_0'(z),$$
$$(\lambda \rho - \lambda z\rho)P_0(z) = (\gamma - z\rho\gamma - z\gamma + \rho z^2 \gamma)P_0'(z),$$
$$\lambda \rho(1 - z)P_0(z) = \gamma(1 - \rho z)(1 - z)P_0'(z).$$

Dividing by $(1 - z)$ and solving for $P_0'(z)$ gives (3.52).

3.42   **(a)** For the parameters in this problem, $\rho = 2/3$ and

$$L_o = \frac{(2/3)^2}{1 - (2/3)} \cdot \frac{15 + 6}{6} = \frac{14}{3}.$$

The wait in orbit for the $M/M/1$ retrial queue is

$$W_o = \frac{L_o}{\lambda} = \frac{7}{15} \text{ hours.}$$

(b) The average rate of calls due to *newly arriving* customers is λ. The average rate of calls due to customers *in orbit* is γL_o. Thus, the total rate of call attempts is $\lambda + \gamma L_o$.

$$\lambda + \gamma L_o = 10 + 6 \cdot \frac{14}{3} = 38 \text{ per hour.}$$

(c) Since every customer *eventually* completes a call, the rate of successful call attempts is λ, the arrival rate of customers. Alternatively, the rate of successful call attempts can be found as:

$$\sum_{n=0}^{\infty}(\lambda + n\gamma)p_{0,n} = \sum_{n=0}^{\infty} \mu p_{1,n} = \mu p_1 = \mu \rho = \lambda.$$

The first equality follows from (3.47). Now, the total rate of call attempts is $\lambda + \gamma L_o$. Thus, the rate of unsuccessful call attempts is $\lambda + \gamma L_o - \lambda = \gamma L_o$. Alternatively, this result can be seen as follows: Every call from the orbit is the result of exactly one previous unsuccessful call attempt; thus, the rate of calls from the orbit is the same as the rate of unsuccessful calls. In summary, the fraction of call attempts that receive a busy signal is

$$\frac{\gamma L_o}{\lambda + \gamma L_o} = \frac{28}{38} \approx 0.737.$$

(d) We have an $M/M/1/1$ queue with $\lambda = 38$ from (b) and $\mu = 15$. The blocking probability for an $M/M/1/1$ queue is

$$\frac{\lambda}{\lambda + \mu} = \frac{38}{53} \approx 0.717.$$

The blocking probability is underestimated using the incorrect $M/M/1/1$ model.

3.45   **(a)** Starting from (3.76):

$$P_1'(1) = \rho\frac{c}{b}\left[(c + 2a - b + 1)\Phi(a + 1, b + 1; c)\right.$$
$$\left. + (b - a)\Phi(a, b + 1; c)\right]p_{0,0}.$$

Write $(c + 2a - b + 1)$ as $(c + a - b + 1) + a$; then apply (3.73) to the term with $(c + a - b + 1)$ and (3.74) to the term with a:

$$(c + 2a - b + 1)\Phi(a + 1, b + 1; c)$$
$$= (c + a - b + 1)\Phi(a + 1, b + 1; c) + a\Phi(a + 1, b + 1; c)$$
$$= \frac{b(c + a - b + 1)}{c}[\Phi(a + 1, b; c) - \Phi(a, b; c)]$$
$$+ (a - b)\Phi(a, b + 1; c) + b\Phi(a, b; c)$$
$$= \frac{b(c + a - b + 1)}{c}\Phi(a + 1, b; c) - \frac{b(a - b + 1)}{c}\Phi(a, b; c)$$
$$+ (a - b)\Phi(a, b + 1; c).$$

Substituting this into the original expression for $P_1'(1)$ gives the desired result:

$$P_1'(1) = \rho[(c + a - b + 1)\Phi(a + 1, b; c) - (a - b + 1)\Phi(a, b; c)]\, p_{0,0}.$$

(b) Starting from (3.78):

$$L_o = [\rho(c + a - b + 1) + a]\Phi(a + 1, b; c)\, p_{0,0}$$
$$- [\rho(a - b + 1) + a]\Phi(a, b; c)\, p_{0,0}.$$

Now

$$\rho(c + a - b + 1) + a$$
$$= \frac{\lambda}{\mu}\left(\frac{q\lambda}{(1 - q)\gamma} + \frac{\lambda}{\gamma} - \frac{\mu + (1 - q)(\lambda + \gamma)}{(1 - q)\gamma} + 1\right) + \frac{\lambda}{\gamma}$$
$$= \frac{\lambda}{\mu\gamma(1 - q)}[q\lambda + \lambda(1 - q) - \mu - (1 - q)(\lambda + \gamma) + (1 - q)\gamma + (1 - q)\mu]$$
$$= \frac{\lambda}{\mu\gamma} \cdot \frac{q}{1 - q} \cdot (\lambda - \mu) = \frac{\lambda}{\mu\gamma} \cdot \frac{q}{1 - q} \cdot (\lambda - \mu) = \frac{\lambda}{\gamma} \cdot \frac{q}{1 - q} \cdot (\rho - 1).$$

Similarly,

$$\rho(a - b + 1) + a = \frac{\lambda}{\mu}\left(\frac{\lambda}{\gamma} - \frac{\mu + (1 - q)(\lambda + \gamma)}{(1 - q)\gamma} + 1\right) + \frac{\lambda}{\gamma}$$
$$= \frac{\lambda}{\mu\gamma(1 - q)}[\lambda(1 - q) - \mu - (1 - q)(\lambda + \gamma) + (1 - q)\gamma + (1 - q)\mu]$$
$$= -\frac{\lambda}{\gamma} \cdot \frac{q}{1 - q}.$$

Plugging these results and $p_{0,0}$ from (3.68) into our original equation for L_o gives:

$$
\begin{aligned}
L_o &= \frac{\lambda}{\gamma} \cdot \frac{q}{1-q} \cdot \frac{(\rho-1)\Phi(a+1,b;c) + \Phi(a,b;c)}{\Phi(a,b;c) + \rho\Phi(a+1,b;c)} \\[2mm]
&= \frac{\lambda}{\gamma} \cdot \frac{q}{1-q} \cdot \frac{1 + (\rho-1)\dfrac{\Phi(a+1,b;c)}{\Phi(a,b;c)}}{1 + \rho\dfrac{\Phi(a+1,b;c)}{\Phi(a,b;c)}} \\[2mm]
&= \frac{\lambda}{\gamma} \cdot \frac{q}{1-q} \cdot \frac{1 + (1-1/\rho)\rho^*}{1 + \rho^*} \\[2mm]
&= \frac{q}{1-q} \cdot \frac{\lambda + (\lambda-\mu)\rho^*}{\gamma(1+\rho^*)}.
\end{aligned}
$$

CHAPTER 4

NETWORKS, SERIES, AND CYCLIC QUEUES

4.3   We use QtsPlus ($M/M/c$ queues-in-series module) with 3 checkout servers. For the initial $M/M/\infty$ queue, we use 150 servers as an approximation for ∞. The total system congestion L increases from 33.4 to 39.0. The queue wait at the counters increases significantly from 1.14 min to 9.57 min (0.1595 h).

M/M/c QUEUES IN SERIES

To setup new problem, enter number of nodes in queueing network. Enter external arrival rate and upper bound on probability range, then press "Solve".

Number of Nodes: 2
External arrival rate: 40.
Maxium no. of probabilities: 5

(Solve)

Node	$1/\mu$	Servers
1	0.75	150
2	0.066667	3

Solutions Manual to Accompany Fundamentals of Queueing Theory, Fourth Edition.   **33**
By D. Gross, J. F. Shortle, J. M. Thompson, and C. M. Harris
Copyright © 2008 John Wiley & Sons, Inc.

Results

Interarrival time (1/λ)		0.025
Expected number in the system (L)		39.04672897
Expected number in all the queues (Lq)		6.380062305
Expected time to clear the system (W)		0.976168224
Expected delay in all the queues (Wq)		0.159501558
Probability that all servers are idle (p0)		2.62363E-15

Node Performance Measures

Node	1	2
Servers	150	3
μ	1.333333333	15
r	30	2.666666667
ρ	20.00%	88.89%
p0	0.00000	0.02804
L	30	9.046728972
Lq	1.8937E-55	6.380062305
W	0.75	0.226168224
Wq	0	0.159501558

Marginal Distributions

0	0.000000	0.028037
1	0.000000	0.074766
2	0.000000	0.099688
3	0.000000	0.088612
4	0.000000	0.078766
5	0.000000	0.070014

4.6   With $\mu_1 = \mu_2 = \mu$, the first equation from (4.7) is

$$p_{0,1} = \frac{\lambda}{\mu} p_{0,0}.$$

The fourth equation gives

$$p_{1,1} = \frac{\lambda}{2\mu} p_{0,1} = \frac{\lambda^2}{2\mu^2} p_{0,0}.$$

The second equation gives

$$p_{1,0} = p_{1,1} + \frac{\lambda}{\mu} p_{0,0} = \left[\frac{\lambda^2}{2\mu^2} + \frac{\lambda}{\mu} \right] p_{0,0} = \frac{\lambda(\lambda + 2\mu)}{2\mu^2} p_{0,0}.$$

The fifth equation gives

$$p_{b,1} = p_{1,1} = \frac{\lambda^2}{2\mu^2} p_{0,0}.$$

The boundary condition gives

$$p_{0,0}^{-1} = 1 + \frac{\lambda}{\mu} + \frac{\lambda^2}{2\mu^2} + \frac{\lambda(\lambda + 2\mu)}{2\mu^2} + \frac{\lambda^2}{2\mu^2}$$

$$= \frac{2\mu^2 + 2\mu\lambda + \lambda^2 + \lambda^2 + 2\mu\lambda + \lambda^2}{2\mu^2} = \frac{3\lambda^2 + 4\mu\lambda + 2\mu^2}{2\mu^2}.$$

4.9   The service rate at each node is $\mu_i \min(n_i, c_i)$. Note that

$$\min(n_i, c_i) = \frac{a_i(n_i)}{a_i(n_{i-1})},$$

where

$$a_i(n_i) = \begin{cases} n_i!, & n_i \leq c_i, \\ c_i^{n_i - c_i} c_i!, & n_i \geq c_i. \end{cases}$$

Equating the flows into and out of state $\bar{n}$ gives the analog of (4.9):

$$\sum_{i=1}^{k} \gamma_i p_{\bar{n};i-} + \sum_{\substack{j=1 \\ (i \neq j)}}^{k} \sum_{i=1}^{k} \frac{\mu_i a_i(n_i + 1)}{a_i(n_i)} r_{ij} p_{\bar{n};i+j-}$$

$$+ \sum_{i=1}^{k} \frac{\mu_i a_i(n_i + 1)}{a_i(n_i)} r_{i0} p_{\bar{n};i+}$$

$$= \sum_{i=1}^{k} \frac{\mu_i (1 - r_{ii}) a_i(n_i)}{a_i(n_i - 1)} p_{\bar{n}} + \sum_{i=1}^{k} \gamma_i p_{\bar{n}}. \qquad (4.9)^*$$

Now we show that (4.12) is a solution to $(4.9)^*$. Equation (4.12) is

$$p_{\bar{n}} = C \frac{r_1^{n_1}}{a_1(n_1)} \cdots \frac{r_i^{n_i}}{a_i(n_i)} \cdots \frac{r_k^{n_k}}{a_k(n_k)},$$

where C is a normalization constant, and $r_i = \lambda_i / \mu_i$ (not to be confused with the routing probabilities r_{ij}). Thus

$$p_{\bar{n};i+1} = C \frac{r_1^{n_1}}{a_1(n_1)} \cdots \frac{r_i^{n_i+1}}{a_i(n_i + 1)} \cdots \frac{r_k^{n_k}}{a_k(n_k)} = p_{\bar{n}} \frac{r_i a_i(n_i)}{a_i(n_i + 1)},$$

and similarly,

$$p_{\bar{n};i-1} = p_{\bar{n}} \frac{a_i(n_i)}{a_i(n_i - 1) r_i},$$

$$p_{\bar{n};i+j-} = p_{\bar{n}} \frac{r_i a_i(n_i) a_j(n_j)}{a_i(n_i + 1) a_j(n_j - 1) r_j}.$$

Substituting into $(4.9)^*$ and canceling the common factor $p_{\bar{n}}$ gives

$$\sum_{i=1}^{k} \gamma_i \frac{a_i(n_i)}{a_i(n_i - 1) r_i} + \sum_{\substack{j=1 \\ (i \neq j)}}^{k} \sum_{i=1}^{k} \frac{\mu_i r_i a_j(n_j)}{a_j(n_j - 1) r_j} r_{ij} + \sum_{i=1}^{k} \mu_i r_{i0} r_i$$

$$\stackrel{?}{=} \sum_{i=1}^{k} \mu_i (1 - r_{ii}) \frac{a_i(n_i)}{a_i(n_i - 1)} + \sum_{i=1}^{k} \gamma_i.$$

Now letting $\bar{a}_i \equiv a_i(n_i)/a_i(n_i - 1)$ gives that

$$\sum_{i=1}^{k} \gamma_i \frac{\mu_i}{\lambda_i} \bar{a}_i + \sum_{\substack{j=1 \\ (i \neq j)}}^{k} \sum_{i=1}^{k} \frac{\mu_j \lambda_i r_{ij}}{\lambda_j} \bar{a}_j + \sum_{i=1}^{k} \lambda_i r_{i0} \overset{?}{=} \sum_{i=1}^{k} \mu_i(1 - r_{ii})\bar{a}_i + \sum_{i=1}^{k} \gamma_i.$$

We move the term with r_{ii} to the double summation and switch the labeling of the indices i and j in the double summation to give

$$\sum_{i=1}^{k} \gamma_i \frac{\mu_i}{\lambda_i} \bar{a}_i + \sum_{j=1}^{k} \sum_{i=1}^{k} \frac{\mu_i \lambda_j r_{ji}}{\lambda_i} \bar{a}_i + \sum_{i=1}^{k} \lambda_i r_{i0} \overset{?}{=} \sum_{i=1}^{k} \mu_i \bar{a}_i + \sum_{i=1}^{k} \gamma_i.$$

Regrouping gives

$$\sum_{i=1}^{k} \left\{ \bar{a}_i \left[\gamma_i \frac{\mu_i}{\lambda_i} + \sum_{j=1}^{k} \frac{\mu_i \lambda_j r_{ji}}{\lambda_i} - \mu_i \right] + \lambda_i r_{i0} \right\} \overset{?}{=} \sum_{i=1}^{k} \gamma_i.$$

Multiplying both sides of (4.10a) by μ_i/λ_i shows that $[\,] = 0$, and hence

$$\sum_{i=1}^{k} \lambda_i r_{i0} \overset{?}{=} \sum_{i=1}^{k} \gamma_i,$$

which is true since the total flow out equals the total flow in.

4.12   The correct model is a two-class two-node open Jackson network where one class (say, class 1) goes first to the chow-mein window and the other class goes first to the rib window. Let node 1 be the chow-mein window and let node 2 be the rib window. Using the QtsPlus two-class open-Jackson-network module, we get the following;

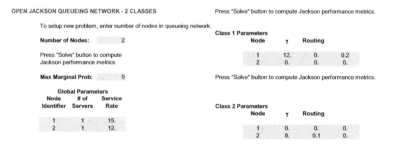

System Performance Measures

Total number in the network (L)	12.318182
Total sojourn time through the network (W)	0.615909

Node Performance Measures

Node	1	2
γ	12.	8.
μ	15.	12.
Servers	1	1
λ	12.8	10.4
ρ	85.33%	86.67%
L	5.818182	6.5
Lq	4.964848	5.633333
W	0.454545	0.625
Wq	0.387879	0.541667

Class 1 Performance Measures

Total number in the network (L)	6.954545
Total sojourn time through the network (W)	0.579545

Class 2 Performance Measures

Total number in the network (L)	5.363636
Total sojourn time through the network (W)	0.670455

Node Performance Measures for Class 1

Node	1	2
γ	12.	0.
μ	15.	12.
Servers	1	1
λ	12	2.4
ρ	80.00%	20.00%
L	5.454545	1.5
Lq	4.654545	1.3
W	0.454545	0.625
Wq	0.387879	0.541667

Node Performance Measures for Class 2

Node	1	2
γ	0.	8.
μ	15.	12.
Servers	1	1
λ	0.8	8
ρ	5.33%	66.67%
L	0.363636	5
Lq	0.310303	4.333333
W	0.454545	0.625
Wq	0.387879	0.541667

Marginal Probabilities

0	0.200000	0.800000
1	0.160000	0.160000
2	0.128000	0.032000
3	0.102400	0.006400
4	0.081920	0.001280
5	0.065536	0.000256

Marginal Probabilities

0	0.946667	0.333333
1	0.050489	0.222222
2	0.002693	0.148148
3	0.000144	0.098765
4	0.000008	0.065844
5	0.000000	0.043896

4.15 The solution is similar to that given in Problem 4.9. First, we modify (4.14) as

$$\sum_{\substack{j=1 \\ (i \neq j)}}^{k} \sum_{i=1}^{k} \frac{\mu_i a_i(n_i + 1)}{a_i(n_i)} r_{ij} p_{\bar{n};i+j-} - \sum_{i=1}^{k} \mu_i (1 - r_{ii}) \frac{a_i(n_i)}{a_i(n_i - 1)} p_{\bar{n}} = 0 .$$

Again, we assume a product form for $p_{\bar{n}}$ as in (4.17) and write

$$p_{\bar{n};i+j-} = p_{\bar{n}} \frac{\rho_i a_i(n_i) a_j(n_j)}{\rho_j a_i(n_i + 1) a_j(n_j - 1)} .$$

Substituting into the previous equation and canceling gives

$$\sum_{\substack{j=1 \\ (i \neq j)}}^{k} \sum_{i=1}^{k} \mu_i r_{ij} \frac{a_j(n_j)}{a_j(n_j - 1)} \frac{\rho_i}{\rho_j} - \sum_{i=1}^{k} \mu_i (1 - r_{ii}) \frac{a_i(n_i)}{a_i(n_i - 1)} \overset{?}{=} 0 .$$

Letting $\bar{a}_i \equiv a_i(n_i)/a_i(n_i - 1)$, we can write as

$$\sum_{\substack{j=1 \\ (i \neq j)}}^{k} \sum_{i=1}^{k} \bar{a}_j \mu_i r_{ij} \frac{\rho_i}{\rho_j} - \sum_{i=1}^{k} \mu_i (1 - r_{ii}) \bar{a}_i \overset{?}{=} 0 .$$

Moving the term with r_{ii} into the double summation and regrouping gives

$$\sum_{j=1}^{k}\left[\sum_{i=1}^{k}\bar{a}_j\mu_i r_{ij}\frac{\rho_i}{\rho_j}-\mu_j\bar{a}_j\right]\stackrel{?}{=}0.$$

Rewriting slightly gives

$$\sum_{j=1}^{k}\bar{a}_j\left[\sum_{i=1}^{k}\mu_i r_{ij}\frac{\rho_i}{\rho_j}-\mu_j\right]\stackrel{?}{=}0.$$

Finally, (4.16) implies that the term in brackets is zero.

4.18   Using the QtsPlus closed-network Buzen algorithm, we have the following:

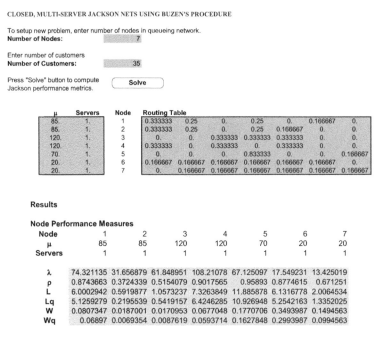

CLOSED, MULTI-SERVER JACKSON NETS USING BUZEN'S PROCEDURE

To setup new problem, enter number of nodes in queueing network.
Number of Nodes: 7

Enter number of customers
Number of Customers: 35

Press "Solve" button to compute   Solve
Jackson performance metrics.

μ	Servers	Node	Routing Table						
85.	1.	1	0.333333	0.25	0.	0.25	0.	0.166667	0.
85.	1.	2	0.333333	0.25	0.	0.25	0.166667	0.	0.
120.	1.	3	0.	0.	0.333333	0.333333	0.333333	0.	0.
120.	1.	4	0.333333	0.	0.333333	0.	0.333333	0.	0.
70.	1.	5	0.	0.	0.	0.	0.833333	0.	0.166667
20.	1.	6	0.166667	0.166667	0.166667	0.166667	0.166667	0.166667	0.
20.	1.	7	0.	0.166667	0.166667	0.166667	0.166667	0.166667	0.166667

Results

Node Performance Measures

Node	1	2	3	4	5	6	7
μ	85	85	120	120	70	20	20
Servers	1	1	1	1	1	1	1
λ	74.321135	31.656879	61.848951	108.21078	67.125097	17.549231	13.425019
ρ	0.8743663	0.3724339	0.5154079	0.9017565	0.95893	0.8774615	0.671251
L	6.0002942	0.5919877	1.0573237	7.3263849	11.885878	6.1316778	2.0064534
Lq	5.1259279	0.2195539	0.5419157	6.4246285	10.926948	5.2542163	1.3352025
W	0.0807347	0.0187001	0.0170953	0.0677048	0.1770706	0.3493987	0.1494563
Wq	0.06897	0.0069354	0.0087619	0.0593714	0.1627848	0.2993987	0.0994563

4.21   Using the QtsPlus closed-Jackson Buzen module, with a transition matrix to make the network a cyclic queue, we get:

CLOSED, MULTI-SERVER JACKSON NETS USING BUZEN'S PROCEDURE

To setup new problem, enter number of nodes in queueing network.
Number of Nodes: 2

Enter number of customers
Number of Customers: 5

Press "Solve" button to compute   Solve
Jackson performance metrics.

μ	Servers	Node	Routing Table	
0.2	2.	repair	0.	1.
0.1	5.	oper mchs	1.	0.

Results

Node Performance Measures   **Marginal Distributions**

Node	repair	oper mchs	Node	repair	oper mchs
μ	0.2	0.1	0	0.11054	0.02591
Servers	2	5	1	0.27634	0.10363
			2	0.27634	0.20725
λ	0.3005181	0.3005181	3	0.20725	0.27634
ρ	0.8894646	0.9740933	4	0.10363	0.27634
L	1.9948187	3.0051813	5	0.02591	0.11054
Lq	0.492228	0			
W	6.637931	10			
Wq	1.637931	0			

For part (a), the probability that exactly one machine is up is .1036. For part (b), W at node 1 is 6.638 so that the performance measure suggested, W divided by the average service time, is $6.638/5 = 1.328$.

For part (c), we have 6 machines in the system (one is a spare), while the maximum number of operating machines remains at 5. Again using the QtsPlus module, the probability that exactly one machine is up is .0864.

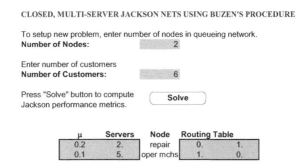

CLOSED, MULTI-SERVER JACKSON NETS USING BUZEN'S PROCEDURE

To setup new problem, enter number of nodes in queueing network.
Number of Nodes: 2

Enter number of customers
Number of Customers: 6

Press "Solve" button to compute   Solve
Jackson performance metrics.

μ	Servers	Node	Routing Table	
0.2	2.	repair	0.	1.
0.1	5.	oper mchs	1.	0.

Results

Node Performance Measures			Marginal Distributions		
Node	repair	oper mchs	**Node**	repair	oper mchs
μ	0.2	0.1	0	0.07375	0.02161
Servers	2	5	1	0.18438	0.08643
			2	0.23048	0.17286
λ	0.3336214	0.3336214	3	0.23048	0.23048
ρ	0.926246	0.9783924	4	0.17286	0.23048
L	2.5900317	3.4099683	5	0.08643	0.18438
Lq	0.9219245	0.073754	6	0.02161	0.07375
W	7.7633851	10.221071			
Wq	2.7633851	0.2210708			

CHAPTER 5

GENERAL ARRIVAL OR SERVICE PATTERNS

5.3 The arrival rate is 18/h = .3/min. From the data, the mean service time is 3.197 min and the variance is 2.514 min^2. Thus, $\rho \doteq .3 \cdot 3.197 \doteq .959$. The expected number in queue and the expected wait in queue are

$$L_q = \frac{\rho^2 + \lambda^2 \sigma_B^2}{2(1-\rho)} \doteq 14.0,$$

$$W_q = \frac{\rho^2/\lambda + \lambda \sigma_B^2}{2(1-\rho)} \doteq 46.7 \text{ min.}$$

5.6 The parameters are $1/\mu = 5.2$ min, $L = 5$, $\lambda = (1/6)/$ min, and $\rho = 13/15$.

$$L = \rho + \frac{\rho^2 + \lambda^2 \sigma_B^2}{2(1-\rho)} \Rightarrow \sigma_B^2 = \frac{2(L-\rho)(1-\rho) - \rho^2}{\lambda^2}.$$

So,

$$\sigma_B^2 = \frac{2(5 - 13/15)(2/15) - 13^2/15^2}{(1/36)} = \frac{79/225}{1/36} = 12.64 \text{ min}^2.$$

Or $\sigma_B \doteq 3.56$ min.

5.9 $\pi_i = \pi_0 k_i + \sum_{j=1}^{i+1} \pi_j k_{i-j+1}$. Multiply by z^i and sum:

$$\sum_{i=0}^{\infty} \pi_i z^i = \pi_0 \sum_{i=0}^{\infty} k_i z^i + \sum_{i=0}^{\infty} \sum_{j=1}^{i+1} \pi_j k_{i-j+1} z^i,$$

or

$$\Pi(z) = \pi_0 K(z) + \sum_{j=1}^{\infty} \sum_{i=j-1}^{\infty} \pi_j k_{i-j+1} z^i$$

$$= \pi_0 K(z) + \frac{1}{z} \sum_{j=1}^{\infty} \pi_j z^j \sum_{i=j-1}^{\infty} k_{i-j+1} z^{i-j+1}$$

$$= \pi_0 K(z) + \frac{(\Pi(z) - \pi_0) K(z)}{z}.$$

Thus

$$\Pi(z) = \frac{\pi_0 K(z)(1 - z)}{K(z) - z}.$$

5.12   In the example, the arrival rate is 5/h = (5/60)/min, the mean service time is 10 min, and the service variance is 2 min^2. Using the QtsPlus $M/G/1$ module, we get the following:

M/G/1: POISSON ARRIVALS TO A SINGLE GENERAL SERVER

Input Parameters:

Arrival rate (λ)	0.083333
Mean service time ($1/\mu$)	10.
Service time variance (σ^2)	2.

Results:

Mean interarrival time ($1/\lambda$)	12.0
Service rate (μ)	0.1
Service time Coefficient of Variation ($\sigma*\mu$)	0.141421
Server utilization (ρ)	83.33%
Fraction of time system is empty (p_0)	0.166667
Mean system size (L)	2.958333
Mean size of queue (Lq)	2.125
Mean system waiting time (W)	35.5
Mean waiting time in the queue (Wq)	25.5
Expected length of the busy period (B)	60.0

5.15   **(a)** The assumption of exponential service times is not justified by the data. For an exponential random variable, the mean and standard deviation are equal. For this data, the sample average transaction time is $748/15 \doteq 49.87$ seconds. The sample standard deviation is about 20.76 seconds, which is less than half of the mean. Therefore an exponential model is not appropriate. The sample variance is computed as follows:

$$\frac{\left(\sum_{i=1}^{n} x_i^2\right) - n\bar{x}^2}{n-1} = \frac{43,332 - 15(748/15)^2}{14} \doteq 430.838.$$

(b) If we measure time in seconds, then $\lambda = 1/60$ per second, $\mu \doteq 1/49.87$ per second, $\sigma_B \doteq 20.76$ seconds, and $\rho \doteq 49.87/60 \doteq 0.831$. Then,

$$L_q = \frac{\rho^2 + \lambda^2 \sigma_B^2}{2(1-\rho)} \doteq 2.34.$$

(c) Solve the following equation for μ:

$$L_q = \frac{\rho^2 + \lambda^2 \sigma_B^2}{2(1-\rho)} = 1.$$

$$\frac{\rho^2 + \lambda^2 \sigma_B^2}{2(1-\rho)} = \frac{\lambda^2 + \lambda^2 \mu^2 \sigma_B^2}{2(\mu^2 - \lambda\mu)} = 1.$$

Thus

$$\lambda^2 + \lambda^2 \mu^2 \sigma_B^2 = 2(\mu^2 - \lambda\mu).$$

Or

$$(2 - \lambda^2 \sigma_B^2)\mu^2 - 2\lambda\mu - \lambda^2 = 0.$$

Using the quadratic formula:

$$\mu = \frac{2\lambda \pm \sqrt{4\lambda^2 + 4\lambda^2(2 - \lambda^2 \sigma_B^2)}}{2(2 - \lambda^2 \sigma_B^2)} = \frac{\lambda \pm \lambda\sqrt{3 - \lambda^2 \sigma_B^2}}{2 - \lambda^2 \sigma_B^2}.$$

So, $\mu \doteq 0.0239$ per second, so the required average transaction time is $1/\mu \doteq 41.8$ seconds.

5.18   For the $M/E_2/1$ queue,

$$B^*(s) = \left(\frac{2\mu}{s + 2\mu}\right)^2.$$

From (5.34),

$$W_q^*(s) = \frac{(1-\rho)s}{s - \lambda + \lambda B^*(s)} = \frac{(1-\rho)s}{s - \lambda + \lambda \left(\dfrac{2\mu}{s + 2\mu}\right)^2}$$

$$= \frac{(1-\rho)s(s+2\mu)^2}{(s-\lambda)(s+2\mu)^2 + 4\lambda\mu^2} = \frac{(1-\rho)(s+2\mu)^2}{s^2 + (4\mu - \lambda)s + 4\mu(\mu - \lambda)}.$$

We use partial fractions to invert

$$W_q^*(s) = \frac{(1-\rho)(s+2\mu)^2}{s^2 + (4\mu - \lambda)s + 4\mu(\mu - \lambda)}$$

$$= (1-\rho) + \frac{(1-\rho)(\lambda s + 4\mu\lambda)}{s^2 + (4\mu - \lambda)s + 4\mu(\mu - \lambda)}$$

$$= (1-\rho) + \frac{C_1}{s - A} + \frac{C_2}{s - B},$$

where

$$A = \frac{\lambda - 4\mu + \sqrt{\lambda^2 + 8\mu\lambda}}{2}, \qquad B = \frac{\lambda - 4\mu - \sqrt{\lambda^2 + 8\mu\lambda}}{2},$$

Completing the partial fractions expansion (after some algebra) gives

$$C_1 = \frac{(1-\rho)\lambda}{2}\left[1 + \frac{\lambda + 4\mu}{\sqrt{\lambda^2 + 8\mu\lambda}}\right],$$

$$C_2 = \frac{(1-\rho)\lambda}{2}\left[1 - \frac{\lambda + 4\mu}{\sqrt{\lambda^2 + 8\mu\lambda}}\right].$$

Rewriting the transform as

$$W_q^*(s) = (1-\rho) + \frac{C_1}{-A} \cdot \frac{-A}{s - A} + \frac{C_2}{-B} \cdot \frac{-B}{s - B}$$

and then inverting gives

$$W_q(t) = (1-\rho) + \frac{-C_1}{A}(1 - e^{At}) + \frac{-C_2}{B}(1 - e^{Bt}).$$

5.21   Taking derivatives of the LST gives

$$G^*(s) = B^*[s + \lambda - \lambda G^*(s)],$$

$$G^{*\prime}(s) = B^{*\prime}[s + \lambda - \lambda G^*(s)] \cdot [1 - \lambda G^{*\prime}(s)],$$

$$G^{*\prime\prime}(s) = B^{*\prime\prime}[s + \lambda - \lambda G^*(s)] \cdot [1 - \lambda G^{*\prime}(s)]^2$$
$$+ B^{*\prime}[s + \lambda - \lambda G^*(s)] \cdot [-\lambda G^{*\prime\prime}(s)].$$

Since $E[X^2] = G^{*\prime\prime}(0)$, we have

$$E[X^2] = B^{*\prime\prime}[\lambda - \lambda G^*(0)] \cdot [1 - \lambda G^{*\prime}(0)]^2$$
$$+ B^{*\prime}[\lambda - \lambda G^*(0)] \cdot [-\lambda G^{*\prime\prime}(0)]$$
$$= B^{*\prime\prime}(0) \cdot (1 + \lambda E[X])^2 + B^{*\prime}(0) \cdot [-\lambda G^{*\prime\prime}(0)]$$
$$= E[S^2] \cdot (1 + \lambda E[X])^2 + (-1/\mu)(-\lambda E[X^2]).$$

Since $E[X] = 1/(\mu - \lambda)$ (as given in the text), this gives

$$E[X^2] = \frac{E[S^2][1 + \lambda/(\mu - \lambda)]^2}{1 - \lambda/\mu} = \frac{E[S^2]}{(1 - \rho)^3}.$$

5.24   Suppose that there are $i \geq 1$ in the system when a customer begins service. The service-time CDF and PDF are

$$B_i(t) = \left\{ \begin{array}{ll} 1 - e^{-\mu_1 t}, & i = 1 \\ 1 - e^{-\mu t}, & i > 1 \end{array} \right. \quad \text{and} \quad b_i(t) = \left\{ \begin{array}{ll} \mu_1 e^{-\mu_1 t}, & i = 1 \\ \mu e^{-\mu t}, & i > 1 \end{array} \right. .$$

The probability of n arrivals during a service time that begins when there are i in the system is

$$k_{n,i} = \int_0^\infty \frac{e^{-\lambda t}(\lambda t)^n \mu e^{-\mu t}}{n!} dt = \frac{\lambda^n \mu}{(\lambda + \mu)^{n+1}} = \frac{\rho^n}{(\rho + 1)^{n+1}}, \quad i > 1,$$

$$k_{n,i} = \int_0^\infty \frac{e^{-\lambda t}(\lambda t)^n \mu_1 e^{-\mu_1 t}}{n!} dt = \frac{\lambda^n \mu_1}{(\lambda + \mu_1)^{n+1}} = \frac{\rho_1^n}{(\rho_1 + 1)^{n+1}}, \quad i = 1.$$

The transition matrix for the embedded Markov chain is

$$P = \begin{bmatrix} \frac{1}{\rho_1+1} & \frac{\rho_1}{(\rho_1+1)^2} & \frac{\rho_1^2}{(\rho_1+1)^3} & \frac{\rho_1^3}{(\rho_1+1)^4} & \cdots \\ \frac{1}{\rho_1+1} & \frac{\rho_1}{(\rho_1+1)^2} & \frac{\rho_1^2}{(\rho_1+1)^3} & \frac{\rho_1^3}{(\rho_1+1)^4} & \cdots \\ 0 & \frac{1}{\rho+1} & \frac{\rho}{(\rho+1)^2} & \frac{\rho^2}{(\rho+1)^3} & \cdots \\ 0 & 0 & \frac{1}{\rho+1} & \frac{\rho}{(\rho+1)^2} & \cdots \\ \vdots & \vdots & \vdots & \vdots & \vdots \end{bmatrix}.$$

To get the steady-state probabilities,

$$\pi_0 = \frac{1}{\rho_1 + 1}\pi_0 + \frac{1}{\rho_1 + 1}\pi_1.$$

So,

$$\pi_1 = \rho_1 \pi_0.$$

$$\pi_1 = \frac{\rho_1}{(\rho_1 + 1)^2}\pi_0 + \frac{\rho_1}{(\rho_1 + 1)^2}\pi_1 + \frac{1}{\rho + 1}\pi_2$$

$$\Rightarrow \pi_2 = (\rho + 1)\left(\rho_1 - \frac{\rho_1}{(\rho_1 + 1)^2} - \frac{\rho_1^2}{(\rho_1 + 1)^2}\right)\pi_0.$$

So,

$$\pi_2 = \left[\frac{\rho + 1}{\rho_1 + 1}\right]\rho_1^2 \pi_0.$$

$$\pi_2 = \frac{\rho_1^2}{(\rho_1+1)^3}\pi_0 + \frac{\rho_1^2}{(\rho_1+1)^3}\pi_1 + \frac{\rho}{(\rho+1)^2}\pi_2 + \frac{1}{(\rho+1)}\pi_3$$

$$\Rightarrow \pi_3 = (\rho+1)\left(\frac{\rho_1^2(\rho+1)}{(\rho_1+1)} - \frac{\rho_1^2}{(\rho_1+1)^3} - \frac{\rho_1^3}{(\rho_1+1)^3}\right.$$
$$\left. - \frac{\rho}{(\rho+1)^2}\frac{\rho_1^2(\rho+1)}{(\rho_1+1)}\right)\pi_0.$$

So,

$$\pi_3 = \left[\frac{\rho^2}{\rho_1+1} + \frac{\rho_1(\rho+1)}{(\rho_1+1)^2}\right]\rho_1^2\pi_0.$$

$$\pi_3 = \frac{\rho_1^3}{(\rho_1+1)^4}\pi_0 + \frac{\rho_1^3}{(\rho_1+1)^4}\pi_1 + \frac{\rho^2}{(\rho+1)^3}\pi_2 + \frac{\rho}{(\rho+1)^2}\pi_3$$
$$+ \frac{1}{(\rho+1)}\pi_4$$

$$\Rightarrow \pi_4 = (\rho+1)\left(\frac{\rho_1^2\rho^2}{\rho_1+1} + \frac{\rho_1^3(\rho+1)}{(\rho_1+1)^2} - \frac{\rho_1^3}{(\rho+1)^4} - \frac{\rho_1^4}{(\rho+1)^4}\right.$$
$$\left. - \frac{\rho^2}{(\rho+1)^3}\frac{\rho_1^2(\rho+1)}{(\rho_1+1)} - \frac{\rho}{(\rho+1)^2}\left(\frac{\rho_1^2\rho^2}{\rho_1+1} + \frac{\rho_1^3(\rho+1)}{(\rho_1+1)^2}\right)\right)\pi_0.$$

So

$$\pi_4 = \left[\frac{\rho^3}{\rho_1+1} + \frac{\rho_1\rho^2}{(\rho_1+1)^2} + \frac{\rho_1^2(\rho+1)}{(\rho_1+1)^3}\right]\rho_1^2\pi_0.$$

A similar pattern continues.

5.27 For any loss system, Little's formula gives that $L = \lambda'W$, where λ' is the arrival rate of unblocked customers. Since $L = p_1 + 2p_2$, it follows that

$$L = p_1 + 2p_2 = \lambda'W = \lambda(1 - p_2)/\mu.$$

Furthermore, combining the equations from Section 5.2.2

$$p_0 = p_0(0,0), \quad p_1 = \frac{p_1(0,0)}{\mu}, \quad p_1(0,0) = \lambda p_0(0,0),$$

gives

$$\lambda p_0 = \mu p_1.$$

Hence we get the following set of simultaneous equations in three unknowns:

$$p_0 + p_1 + p_2 = 1,$$
$$\lambda p_0 - \mu p_1 = 0,$$
$$\mu p_1 + (2\mu + \lambda)p_2 = \lambda,$$

which solves to

$$p_1 = \frac{\lambda}{\mu} p_0 \,,$$

$$p_2 = \frac{\lambda^2}{2\mu^2} p_0 \,,$$

$$p_0 = \frac{1}{1 + \frac{\lambda}{\mu} + \frac{\lambda^2}{2\mu^2}} = \frac{2\mu^2}{2\mu^2 + 2\lambda\mu + \lambda^2} \,.$$

5.30 The steady-state probabilities satisfy the relation

$$p_n = p_0 \prod_{i=1}^{n} \frac{\lambda_{i-1}}{\mu_i} \qquad (n = 0, 1, \ldots, c).$$

where

$$\lambda_j = (M - j)\lambda \qquad (j = 0, 1, \ldots, c - 1),$$
$$\mu_i = i\mu \qquad (i = 1, 2, \ldots, c).$$

This implies that

$$p_n = p_0 \prod_{i=1}^{n} \frac{(M - i + 1)\lambda}{i\mu} = p_0 \binom{M}{n} \left(\frac{\lambda}{\mu}\right)^n .$$

The normalizing condition $\sum_{n=0}^{c} p_n = 1$ gives the value for p_0.

5.33 For the $M/M/1$ system, $A^*(s) = \lambda/(\lambda + s)$, so

$$\beta(z) = A^*[\mu(1 - z)] = \frac{\lambda}{\lambda + \mu(1 - z)}.$$

Thus $\beta(z) = z$ is

$$\frac{\lambda}{\lambda + \mu(1 - z)} = z$$

or

$$z^2 - (1 + \rho)z + \rho = 0 = (z - 1)(z - \rho)$$

Thus the root in $(0, 1)$ is $r_0 = \rho$. Also, $q_n = (1 - \rho)\rho^n$ is the $M/M/1$ result.

5.36   Using successive substitution starting from 0.4 gives

z	Righthand Side
0.400000	0.473358
0.473358	0.496795
0.496795	0.505522
0.505522	0.508945
0.508945	0.510313
0.510313	0.510865
0.510865	0.511088
0.511088	0.511178
0.511178	0.511214
0.511214	0.511229
0.511229	0.511235

Thus the required root is 0.511 and the line delay CDF is given by

$$W_q(t) = 1 - 0.511 \exp^{-(1-0.511)t} = 1 - 0.511 \exp^{-0.498t}.$$

5.39   The Markov chain is the same as in (5.50) and (5.51), but with

$$b_n = \int_0^\infty \frac{e^{-(r+\mu)t}[(r+\mu)t]^n}{n!} \, dA(t).$$

Effectively, the queue behaves like a $G/M/1$ queue with service rate $r + \mu$. (The assumption that the reneging rate does not depend on the number in the system is a questionable assumption.)

5.42   We use the $H/M/c$ (hyper or mixed-exponential arrival) module from Qt-sPlus.

H/M/c (MIXED): HYPEREXPONENTIAL INPUT, MULTIPLE EXPONENTIAL SERVERS/UNLIMITED QUEUE
Inter-arrival density function a(t) = sumof(i=1 to n, q(i)*lambda(i)*exp(-lambda(i)*t))
where sumof(i=1 to n, q(i)) = 1

Input Parameters:		q(i)	lambda(i)
Number of alternate phases (n)	2	0.666667	0.5
Mean time to complete service (1/μ)	9.	0.333333	0.25
Number of servers	4		
Total time horizon for prob plotting (T)	25.		
Maximum value of variable whose probability is to be plotted (K)	10		

Solve

Results for H/M/c (Mixed):

Arrival rate [arrivals/unit of time] (l)	0.375
Mean interarrival time (1/l)	2.666667
Service rate[# served/unit of time] (m)	0.111111
Number of servers (c)	4
Fraction of time the server is busy [MUST BE < 1] (r)	0.84375
Function root[determines answers] (r)	0.85862
Fraction of time arrival finds server idle (q0)	0.020025
Probability arrival finds c customers (qc)	0.099001
Expected system size (L)	**7.554014**
Expected system size seen by an arrival (L(A))	7.666176
Expected queue size (Lq)	4.179014
Expected queue size seen by an arrival (Lq(A))	4.252661
Expected waiting time in the system (W)	**20.14404**
Expected waiting time in the queue (Wq)	11.14404

CHAPTER 6

GENERAL MODELS AND THEORETICAL TOPICS

6.3

$$\overline{W}(s) = \frac{1}{s} - \frac{3s^2 + 22s + 36}{3(s^2 + 6s + 8)(s + 3)} = \frac{1}{s} - \left[\frac{2/3}{s+2} + \frac{1}{s+3} - \frac{2/3}{s+4} \right].$$

Thus

$$W(t) = 1 - \frac{2}{3}e^{-2t} - e^{-3t} + \frac{2}{3}e^{-4t}, \quad \text{and} \quad W = \frac{2}{3} \cdot \frac{1}{2} + \frac{1}{3} - \frac{2}{3} \cdot \frac{1}{4} = \frac{1}{2}.$$

6.6 The wait in queue is the sum of 11 $U(0, 1)$ values plus a residual service time. Since the server is busy, this remaining service time has mean (see Problem 5.7)

$$\frac{\mathrm{E}[S^2]}{2\mathrm{E}[S]} = \frac{1/3}{2/2} = \frac{1}{3}.$$

On average, the 11 customers require 11/2 for their service. Thus

$$\mathrm{E}[\text{wait in queue}] = \frac{11}{2} + \frac{1}{3} = \frac{35}{6}.$$

6.9   Multiplying the ith equation by z^i and then summing on i gives

$$P(z) = P_c e^{-\lambda} \sum_{i=0}^{\infty} \frac{\lambda^i}{i!} z^i + p_{c+1} e^{-\lambda} \sum_{i=1}^{\infty} \frac{\lambda^{i-1}}{(i-1)!} z^i$$

$$+ p_{c+2} e^{-\lambda} \sum_{i=2}^{\infty} \frac{\lambda^{i-2}}{(i-2)!} z^i + \sum_{j=3}^{\infty} p_{c+j} e^{-\lambda} \sum_{i=j}^{\infty} \frac{\lambda^{i-j}}{(i-j)!} z^i + \cdots$$

$$= P_c e^{-\lambda} e^{\lambda z} + p_{c+1} z e^{-\lambda} e^{\lambda z} + p_{c+2} z^2 e^{-\lambda} e^{\lambda z} + \sum_{i=3}^{\infty} p_{c+j} z^j e^{-\lambda} e^{\lambda z}$$

$$= P_c e^{-\lambda(1-z)} + \sum_{j=1}^{\infty} p_{c+j} z^j e^{-\lambda(1-z)} .$$

But

$$\sum_{j=1}^{\infty} p_{c+j} z^{c+j} / z^c = \left[P(z) - \sum_{j=0}^{c} p_j z^j \right] / z^c.$$

Thus

$$P(z) = e^{-\lambda(1-z)} \left\{ P_c + \left[P(z) - \sum_{j=0}^{c} p_j z^j \right] / z^c \right\},$$

$$[1 - e^{-\lambda(1-z)} / z^c] P(z) = e^{-\lambda(1-z)} \left[P_c - \sum_{j=0}^{c} p_j z^j / z^c \right],$$

$$P(z) = \frac{\left(\sum_{j=0}^{c} p_j z^j \right) - P_c z^c}{1 - z^c e^{\lambda(1-z)}}.$$

6.12   Label the system states as $0, 1, 2, \bar{2}, 3, \bar{3}, 4, \bar{4}, \ldots$, where n means that there are n in the system and 1 in service, and $\bar{n}$ means that there are n in the system and 2 in service. Let c_j denote the probability that a batch arrival contains j units $(j = 1, 2, \ldots)$. Then,

$$p_{01} = c_1, \qquad\qquad\qquad p_{0\bar{j}} = c_j, \qquad\qquad\qquad j \geq 2,$$

$$p_{i,i-1} = \left(\frac{\mu}{\lambda + \mu}\right), \quad i = 1, 2, \quad p_{i,\overline{i-1}} = \left(\frac{\mu}{\mu + \lambda}\right), \quad i \geq 3,$$

$$p_{i,i+j} = \left(\frac{\lambda}{\lambda + \mu}\right) c_j, \quad i \geq 1, \quad p_{\bar{i},\overline{i+j}} = \left(\frac{\lambda}{\lambda + \mu}\right) c_j, \quad i \geq 2,$$

$$p_{\bar{i},i-2} = \left(\frac{\mu}{\lambda + \mu}\right), \quad i = 2, 3, \quad p_{\bar{i},\overline{i-2}} = \left(\frac{\mu}{\lambda + \mu}\right), \quad i \geq 4.$$

$$Q_{01}(t) = \Pr\{\text{arrival of size 1 occurs}\}\, \Pr\{\text{arrival occurs by } t\}$$
$$= c_1(1 - e^{-\lambda t})$$
$$Q_{0,\bar{j}}(t) = \Pr\{\text{arrival of size } j \text{ occurs}\}\, \Pr\{\text{arrival occurs by } t\}$$
$$= c_j(1 - e^{-\lambda t}) \quad \text{for } j \geq 2$$
$$Q_{i,i+j}(t) = \Pr\{\text{a transition occurs by } t\}\, \Pr\{\text{transition is an arrival}\}$$
$$\times \Pr\{\text{arrival size } j\}$$
$$= [1 - e^{-(\lambda+\mu)t}]\left(\frac{\lambda}{\lambda + \mu}\right) c_j$$
$$Q_{i,i-1}(t) = \Pr\{\text{a transition occurs by } t\}\, \Pr\{\text{transition is a service}\}$$
$$= [1 - e^{-(\lambda+\mu)t}]\left(\frac{\mu}{\lambda + \mu}\right) \quad \text{for } i = 1, 2$$
$$Q_{i,\overline{i-1}}(t) = \Pr\{\text{a transition occurs by } t\}\, \Pr\{\text{transition is a service}\}$$
$$= [1 - e^{-(\lambda+\mu)t}]\left(\frac{\mu}{\lambda + \mu}\right) \quad \text{for } i > 2$$
$$Q_{\bar{i},\overline{i+j}}(t) = \Pr\{\text{a transition occurs by } t\}\, \Pr\{\text{transition is an arrival}\}$$
$$\times \Pr\{\text{arrival size } j\}$$
$$= [1 - e^{-(\lambda+\mu)t}]\left(\frac{\lambda}{\lambda + \mu}\right) c_j \quad \text{for } i \geq 2$$
$$Q_{\bar{i},i-2}(t) = \Pr\{\text{a transition occurs by } t\}\, \Pr\{\text{transition is a service}\}$$
$$= [1 - e^{-(\lambda+\mu)t}]\left(\frac{\mu}{\lambda + \mu}\right) \quad \text{for } i = 2, 3$$
$$Q_{\bar{i},\overline{i-2}}(t) = \Pr\{\text{a transition occurs by } t\}\, \Pr\{\text{transition is a service}\}$$
$$= [1 - e^{-(\lambda+\mu)t}]\left(\frac{\mu}{\lambda + \mu}\right) \quad \text{for } i > 3$$

$$p_{01} = c_1$$
$$p_{0\bar{j}} = c_j, \ j \geq 2$$
$$p_{i,i+j} = \left(\frac{\lambda}{\lambda+\mu}\right) c_j$$
$$p_{i,i-1} = \left(\frac{\mu}{\lambda+\mu}\right), \ i = 1, 2$$
$$p_{i,\overline{i-1}} = \left(\frac{\mu}{\mu+\lambda}\right), \ i > 2$$
$$p_{\bar{i},\overline{i+j}} = \left(\frac{\lambda}{\lambda+\mu}\right) c_j, \ j \geq 2$$
$$p_{\bar{i},i-2} = \left(\frac{\mu}{\lambda+\mu}\right), \ i = 2, 3$$
$$p_{\bar{i},\overline{i-2}} = \left(\frac{\mu}{\lambda+\mu}\right), \ i > 3$$
$$G_0(t) = \sum_{j=0}^{\infty} Q_{0j}(t) = 1 - e^{-\lambda t},$$

$$F_{0,1}(t) = 1 - e^{-\lambda t}$$
$$F_{0,\bar{j}}(t) = 1 - e^{-\lambda t}$$
$$F_{i,i+j}(t) = 1 - e^{-(\lambda+\mu)t}$$
$$F_{i,i-1}(t) = 1 - e^{-(\lambda+\mu)t}, \ i = 1, 2$$
$$F_{i,\overline{i-1}}(t) = 1 - e^{-(\lambda+\mu)t}, \ i > 2$$
$$F_{\bar{i},\overline{i-1}}(t) = 1 - e^{-(\lambda+\mu)t}, \ i \geq 2$$
$$F_{\bar{i},i-2}(t) = 1 - e^{-(\lambda+\mu)t}, \ i = 2, 3$$
$$F_{\bar{i},\overline{i-2}}(t) = 1 - e^{-(\lambda+\mu)t}, \ i > 3$$
$$G_i(t) = 1 - e^{-(\lambda+\mu)t}, \ i \geq 1$$
$$G_{\bar{i}}(t) = 1 - e^{-(\lambda+\mu)t}, \ i \geq 2$$

6.15   This is a $G/M/1$ model, where $G = D_3$. Using QtsPlus we get the following:

G/M/1: GENERAL INPUT, EXPONENTIAL SERVICE, SINGLE-SERVER QUEUE

To start a new problem, enter number of interarrival points and probabilities to be specified.

Number of interarrival probabilities: 3

Enter interarrival time probability distribution data below and then press the "Solve" button.

Input Parameters:

Mean time to complete service (1/μ)	2.
Total time horizon for plotting (T)	20.
Maximum system size for plotting (K)	10

Solve

Interarrival Time (t)	Probability a(t)
2.	0.2
3.	0.7
4.	0.1

Results for G/M/1

Size	General-time p(n)	CDF p(n)
0	0.310345	0.310345
1	0.368605	0.678950
2	0.171594	0.850544
3	0.079881	0.930425
4	0.037186	0.967611
5	0.017311	0.984922
6	0.008059	0.992981
7	0.003752	0.996732
8	0.001746	0.998479
9	0.000813	0.999292
10	0.000378	0.999670

Results for G/M/1:

Arrival rate [arrivals/unit of time] (l)	0.344828
Mean interarrival time (1/l)	2.9
Service rate[# served/unit of time](m)	0.5
Fraction of time the server is busy [MUST BE < 1](r)	0.689655
Function root[determines answers] (r)	0.465523
Fraction of time the server is idle (p0)	0.310345
Expected system size (L)	1.290337
Expected sysems size seen by an arrival (L(A))	0.870988
Expected queue size (Lq)	0.600682
Expected queue size seen by an arrival (Lq(A))	0.405465
Expected non-empty queue size (L'q)	1.870988
Expected waiting time in the system (W)	3.741977
Expected waiting time in the queue (Wq)	1.741977

6.18   This is a $G/M/2$ model, where $G = D_3$, and from QtsPlus we get the following:

G/M/c: GENERAL INPUT, MULTIPLE EXPONENTIAL SERVERS/UNLIMITED QUEUE

To start a new problem, enter number of interarrival
points and probabilities to be specified.
**Number of interarrival
probabilities:**   3

Enter interarrival time probability distribution
data below and then press the "Solve" button.

Input Parameters:

Mean time to complete service (1/μ)	4.
Number of servers (c)	2
Total time horizon for plotting (T)	10.
Maximum system size for plotting (K)	10

Solve

Interarrival Time (t)	Probability a(t)
2.	0.2
3.	0.7
4.	0.1

Results for G/M/c:

			Results for G/M/c General-time	
Arrival rate [arrivals/unit of time] (l)	0.344828	Size	p(n)	CDF p(n)
Mean interarrival time (1/l)	2.9	0	0.128980	0.128980
Service rate[# served/unit of time] (m)	0.25	1	0.362729	0.491709
Number of servers (c)	2	2	0.271670	0.763379
Fraction of time the server is busy [MUST BE < 1] (r)	0.689655	3	0.126468	0.889847
Function root[determines answers] (r)	0.465523	4	0.058874	0.948721
Fraction of time arrival finds server idle (q0)	0.262979	5	0.027407	0.976129
Probability arrival finds c customers (qc)	0.183379	6	0.012759	0.988887
Expected system size (L)	1.822025	7	0.005939	0.994827
Expected system size seen by an arrival (L(A))	1.378958	8	0.002765	0.997592
Expected queue size (Lq)	0.442715	9	0.001287	0.998879
Expected queue size seen by an arrival (Lq(A))	0.298836	10	0.000599	0.999478
Expected waiting time in the system (W)	5.283874			
Expected waiting time in the queue (Wq)	1.283874			

6.21   The solution is the same as in Problem 2.51 but with $C_1 = \$24$, $C_2 = \$138$, and $C_3 = \$10$. That is, we compute

$$E[C] = \text{total costs/h} = C_1 \sum_{n=1}^{k-1} p_n + C_2 \left(1 - \sum_{n=0}^{k-1} p_n \right) + C_3 L$$

for various values of k. When $k = 1$, we have

$$p_0 = \frac{1}{3}, \quad L = 2,$$
$$E[C(1)] = 138(1 - 1/3) + 10(2) = \$112.$$

When $k = 2$, we have

$$p_0 = \frac{1}{5}, \quad p_1 = \frac{4}{15}, \quad L = 2.4,$$
$$E[C(2)] = 24(4/15) + 138(1 - 7/15) + 10(2.4) = \$104.$$

When $k = 3$, we have

$$p_0 = .13, \quad p_1 + p_2 = .41, \quad p_0 + p_1 + p_2 = .54, \quad L = 2.96,$$
$$E[C(3)] = 24(.41) + 138(1 - .54) + 10(2.96) = \$102.92.$$

When $k = 4$, we have

$$p_0 = .09, \quad \sum_{n=1}^{3} p_n = .49, \quad \sum_{n=0}^{3} p_n = .58, \quad L = 3.6,$$
$$E[C(4)] = 24(.49) + 138(1 - .58) + 10(3.6) = \$105.72.$$

The optimal value is $k = 3$.

6.24 Using results for a birth–death process (2.3) gives

$$p_n = p_0 \frac{\lambda e^{-(n-1)/\mu} \cdot \lambda e^{-(n-2)/\mu} \cdots \lambda e^{-1/\mu} \cdot \lambda e^{0/\mu}}{\mu^n}$$
$$= p_0 \rho^n \exp\left[-\sum_{i=0}^{n-1} \frac{i}{\mu}\right]$$
$$= p_0 \rho^n \exp\left[-\frac{(n^2 - n)}{2\mu}\right],$$

and

$$p_0 = \left(\sum_{n=0}^{\infty} \rho^n \exp\left[-\frac{(n^2 - n)}{2\mu}\right]\right)^{-1}.$$

The expected hourly cost is

$$E[C(\mu)] = \$1.50\mu + \$75 \cdot E[\text{customers lost per hour}].$$

The expected number of customers lost per hour is $\sum_n (1 - b_n) p_n \lambda$. Thus

$$E[C(\mu)] = 1.50\mu + 75 \cdot 10 \cdot p_0 \sum_{n=0}^{\infty} (1 - e^{-n/\mu}) \left(\frac{10}{\mu}\right)^n e^{-n(n-1)/2\mu}$$
$$= 1.50\mu + 750 - 750 p_0 \sum_{n=0}^{\infty} \left(\frac{10}{\mu}\right)^n e^{-n(n+1)/2\mu}.$$

A computer program or spreadsheet can be written to estimate p_0 and p_n for fixed values of μ. Then a search procedure can be used to locate the value of μ that minimizes $E[C(\mu)]$. This is found to be approximately $\mu = 26$ with $E[C(\mu)] \doteq \$55$.

6.27 From the first equation, we have

$$\hat{\mu} = \hat{\lambda} + \frac{\hat{\lambda}}{n_a + n_0 - \hat{\lambda} t}.$$

Plugging this into the second equation gives

$$0 = -t_b + \frac{n_c - n_0}{\hat{\lambda} + \frac{\hat{\lambda}}{n_a + n_0 - \hat{\lambda}t}} + \frac{n_a + n_0 - \hat{\lambda}t}{\hat{\lambda} + \frac{\hat{\lambda}}{n_a + n_0 - \hat{\lambda}t}}.$$

After some algebra, this becomes

$$\hat{\lambda}t_b(1 + n_a + n_0 - \hat{\lambda}t) = (n_c + n_a - \hat{\lambda}t)(n_a + n_0 - \hat{\lambda}t),$$

or

$$\underbrace{(t^2 + tt_b)}_{a} \hat{\lambda}^2 - \underbrace{[(n_c + 2n_a + n_0)t + (n_a + n_0 + 1)t_b]}_{-b} \hat{\lambda}$$

$$+ \underbrace{(n_c + n_a)(n_a + n_0)}_{c} = 0,$$

which implies that

$$\hat{\lambda} = \frac{-b \pm \sqrt{b^2 - 4ac}}{2a}.$$

6.30   From the equations following (6.35), the upper confidence limit is

$$\rho_u = \frac{n_c \hat{\rho}}{n_a F_{2n_c, 2n_a}(\alpha/2)} = \frac{8 \cdot \frac{4}{3}}{16 F_{16,32}(.025)} \doteq 1.702.$$

The lower confidence limit is

$$\rho_l = \frac{n_c \hat{\rho}}{n_a F_{2n_c, 2n_a}(1 - \alpha/2)} = \frac{8 \cdot \frac{4}{3}}{16 F_{16,32}(.975)} \doteq 0.297.$$

6.33   For the $M/G/1$ queue,

$$W = \frac{1}{\mu} + \frac{(\lambda/\mu)^2 + \lambda^2 \sigma_B^2}{2\lambda(1 - \lambda/\mu)}.$$

When the service distribution is E_2, then $\sigma_B^2 = 1/2\mu^2$, so

$$W = \frac{1}{\mu} + \frac{(\lambda/\mu)^2 + \lambda^2/2\mu^2}{2\lambda(1 - \lambda/\mu)} = \frac{4\mu - \lambda}{4\mu(\mu - \lambda)}.$$

This implies that $\hat{\mu}$ is the solution to

$$4\hat{W}\hat{\mu}^2 - 4\hat{W}\hat{\lambda}\hat{\mu} - 4\hat{\mu} + \hat{\lambda} = 0.$$

CHAPTER 7

BOUNDS AND APPROXIMATIONS

7.3 Any $D/D/1$ queue with $\rho < 1$ works. For such a queue, $\sigma_A^2 = \sigma_B^2 = 0$, so (7.13) gives that $W_q \leq 0$. In fact, $W_q = 0$, since every interarrival time is greater than every service time.

7.6 Let $W_q^{(n)}$ be the queue delay of the nth customer. $W_q^{(n)}$ is a discrete-time Markov chain where

$$W_q^{(n+1)} = \max(0, W_q^{(n)} + S^{(n)} - T^{(n)}).$$

The possible values of $U^{(n)} = S^{(n)} - T^{(n)}$ are $(-2, 1)$ with probabilities $(\frac{1}{2}, \frac{1}{2})$. So the transition probabilities of this Markov chain are

$$p_{i,i-2} = p_{i,i+1} = 1/2 \quad (i \geq 2),$$
$$p_{00} = p_{01} = 1/2,$$
$$p_{10} = p_{12} = 1/2.$$

Let w_n be the steady-state solution to this Markov chain. That is, w_n is the probability that the queue delay of a customer in steady state is n time units. Then

$$w_0 = w_0 p_{00} + w_1 p_{10} + w_2 p_{20} = \tfrac{1}{2}(w_0 + w_1 + w_2)$$

and

$$w_n = \tfrac{1}{2}w_{n-1} + \tfrac{1}{2}w_{n+2} \quad (n \geq 1).$$

The operator equation is

$$D^3 - 2D + 1 = 0,$$

which has roots $1, (-1 \pm \sqrt{5})/2$. Therefore

$$w_n = C\left(\frac{\sqrt{5}-1}{2}\right)^n \quad (n \geq 0).$$

The normalizing condition that the probabilities sum to 1 gives

$$C = 1 - \frac{\sqrt{5}-1}{2} = \frac{3 - \sqrt{5}}{2}.$$

7.9   The parameters for this problem are $\lambda = 20/\text{h}$, $\mu = 15/\text{h}$, and $\rho = \tfrac{4}{3}$. The system is saturated. Using the equation after (7.26) we get

$$\Pr\{W_q^{(n)} = 0\} = \int_{-\infty}^{\alpha\sqrt{n}/\beta} \frac{e^{-t^2/2}}{\sqrt{2\pi}}\,dt,$$

where

$$\alpha = \mathrm{E}[T] - \mathrm{E}[S] = 3 - 4 = -1\,\text{min},$$
$$\beta^2 = \mathrm{Var}[S] + \mathrm{Var}[T] = 16 + 9 = 25\,\text{min}^2.$$

Thus

$$\Pr\{W_q^{(n)} = 0\} = \int_{-\infty}^{-\sqrt{n}/5} \frac{e^{-t^2/2}}{\sqrt{2\pi}}\,dt.$$

Using Chebyshev's inequality,

$$\Pr\{W_q^{(n)} = 0\} \leq \frac{25}{2n}.$$

The following table shows a comparison in estimating

$$\Pr\{W_q^{(n)} > 0\} = 1 - \Pr\{W_q^{(n)} = 0\}.$$

	Probability of nth Customer Waiting	
n	Normal Approx.	Chebyshev's Ineq.
50	.921	$\geq .75$
100	.977	$\geq .875$
250	.999	$\geq .95$
500	1.000	$\geq .975$
1000	1.000	$\geq .9875$
5000	1.000	$\geq .9975$

CHAPTER 8

NUMERICAL TECHNIQUES AND SIMULATION

8.3   First we give an intuitive derivation by relating the Erlang distribution to a Poisson process. Consider a Poisson process with rate θ. The time until the kth arrival is the sum of k exponential random variables, each with mean $1/\theta$. This is the same as an Erlang random variable with mean k/θ and variance k/θ^2 (which is the Erlang distribution described in the problem statement). The probability that the kth arrival is after t is the same as the probability that there are $k - 1$ or fewer arrivals by t. So the complementary CDF of the Erlang distribution is

$$F^c(t) = \sum_{i=0}^{k-1} \frac{(\theta t)^i e^{-\theta t}}{i!}.$$

Thus

$$F(t) = 1 - F^c(t) = 1 - \sum_{i=0}^{k-1} \frac{(\theta t)^i e^{-\theta t}}{i!},$$

which is (8.36). Alternatively, a formal derivation is given in the text; see (1.15) and the subsequent equations. Equation (1.15) gives the complementary CDF of an Erlang distribution that is the sum of $n + 1$ exponential random variables, each with mean $1/\lambda$. Following the derivation after (1.15), but replacing n with $k-1$ and λ with θ, gives (8.36).

8.6   **(a)** The basic strategy is to first generate a uniform random number to determine whether the final number is drawn from the first exponential distribution (with probability $\frac{1}{3}$) or from the second exponential distribution (with probability $\frac{2}{3}$). The particular exponential random variable is then drawn using the standard inversion method. A sample spreadsheet formula in Excel that does this is

```
=IF(RAND()<1/3, -5 * LN(RAND()), -10 * LN(RAND())).
```

(b) For the gamma distribution, the CDF cannot be inverted analytically. However, Excel provides a function GAMMAINV that computes the inversion numerically. That is, given a probability p, it returns the value x such that $F(x) = p$. The mean of a gamma random variable is $\alpha\beta$, and the variance is $\alpha\beta^2$. Since the mean and variance are 5 and 10, this implies that $\beta = 2$ and $\alpha = 2.5$. A sample spreadsheet formula that generates such a gamma random variable is

```
=GAMMAINV(RAND(), 2.5, 2).
```

8.9   A simulation program written in the JAVA language illustrates how this problem can be solved. The parameters for the simulation are Poisson arrival rate of 0.9 and exponential service time with mean 0.75. We simulate 5×10^6 transactions.

```
/*
 *
 */

public class M_M_1QueueSimulator {
    static final double INFINITY = 1e308;
    static final double IAT_PARM = 1/0.9;        //mean
        inter-arrival time
    static final double ST_PARM = 0.75;          //mean
        service time

    public static void main (String[] args) {
        double CurrentClock;        //Main simulation
            clock
        double TimeForNextArrival;    //Time next
            transaction arrives
        double TimeForServiceCompletion;    //Time for
            service completion
        long MaxDepartures;          //Maximum number of
            simulation events to generate
        long i;                      //working index variables
```

```
long SystemSize;              //Number of
   transactions in the system
double AvgSystemSize;         //Average number
   of transactions in system
double ServerBusyTime;        //Time the server
    is busy
double SystemSizeAccumulator; //accumulator
   for time averaged queue size
long NumberOfArrivals;        //Number of car
   arrivals to ramp
long NumberOfDepartures;      //Number of car
   departures from ramp
double ArrivalRate;           //Average arrival
   rate of cars
double AverageWaitTime;       //Average Wait
   Time
double AverageServerBusy;     //Fraction of
   time server is busy
double ServiceTime;           //Service time for
   the transaction


/*initialize simulation parameters*/
MaxDepartures = 5000000;
NumberOfArrivals = 0;
NumberOfDepartures = 0;
SystemSizeAccumulator = 0;
SystemSize = 0;
ServerBusyTime = 0;
TimeForServiceCompletion = INFINITY;
TimeForNextArrival = iat();
CurrentClock = 0;
AverageWaitTime = 0;


// Do main simulation loop
while (NumberOfDepartures < MaxDepartures){
  if (TimeForNextArrival <
    TimeForServiceCompletion) {
    //process an arrival
    SystemSizeAccumulator =
       SystemSizeAccumulator +
      SystemSize*(TimeForNextArrival −
         CurrentClock);
    SystemSize = SystemSize + 1;
```

```java
        NumberOfArrivals = NumberOfArrivals + 1;
        CurrentClock = TimeForNextArrival;
        TimeForNextArrival = CurrentClock + iat()
           ;

        if (SystemSize == 1){
          ServiceTime = st();
          ServerBusyTime += ServiceTime;
          TimeForServiceCompletion = CurrentClock
              + ServiceTime;
        }
      } else {
        //process a service completion
        SystemSizeAccumulator =
            SystemSizeAccumulator +
          SystemSize*(TimeForServiceCompletion -
              CurrentClock);
        SystemSize = SystemSize - 1;
        NumberOfDepartures = NumberOfDepartures +
            1;
        CurrentClock = TimeForServiceCompletion;

        if (SystemSize > 0) {
          ServiceTime = st();
          ServerBusyTime += ServiceTime;
          TimeForServiceCompletion = CurrentClock
              + ServiceTime;
        } else {
          TimeForServiceCompletion = INFINITY;
        }
      }
    }

  //calculate simulation results
  AvgSystemSize = SystemSizeAccumulator /
      CurrentClock;
  AverageServerBusy = ServerBusyTime/CurrentClock
      ;
  ArrivalRate = NumberOfArrivals / CurrentClock;
  AverageWaitTime = AvgSystemSize / ArrivalRate;
      //using Little's Law

    System.out.println("Percent server is busy (
        rho):");
    System.out.println(100*AverageServerBusy);
```

```java
    System.out.println("Average number in system (L
        ): ");
      System.out.println(AvgSystemSize);
      System.out.println("Avverage number in queue
          (Lq):");
      System.out.println(AvgSystemSize -
          AverageServerBusy);
      System.out.println("Average wait time (W): ")
          ;
      System.out.println(AverageWaitTime);
      System.out.println("Average time in queue (Wq
          ):");
      System.out.println(AverageWaitTime-ST_PARM);
      System.out.println("Number of transactions at
          simulation end:");
      System.out.println(SystemSize);
      System.out.println("Number of arrivals:");
      System.out.println(NumberOfArrivals);
      System.out.println("Number of departures:");
      System.out.println(NumberOfDepartures);
      System.out.println("Simulation clock at end:"
          );
    System.out.println(CurrentClock);
    }

    private static double iat(){
    //generate inter-arrival time expoential with
        mean of 1

      return -IAT_PARM*Math.log(Math.random());
    }

    private static double st(){
    //generate service time exponential with mean
        0.75
      return -ST_PARM*Math.log(Math.random());
    }
}
```

As sample output, the program estimates the average number in the system L as 2.081 and the average wait W as 2.310. Using the analytic $M/M/1$ model in QtsPlus, $L \doteq 2.077$ and $W \doteq 2.308$.

8.12   To illustrate the variety of solution possibilities, this problem is solved with two different programming languages. The first program is written in the C programming language. The C version takes advantage of the fact that the mean time to failure and the mean time to repair are the same.

```c
#include <stdio.h>
#include <stdlib.h>
#include <math.h>
main(){
    /*
        The following are key variables used in
            this program
        k represents the run length
        ct1 and ct reprsents clock counters
        al is the accumulator for mean number of
            machines in repair
        n is the number of machines down
        m is the number of operating machines
            Note:  m + n = 4 at all times
        nta is the number of arrivals to the
            service facility
        ntd is the number of repair completions

        This program takes advantage that the mean-
            time-between-failures
        and mean-time-to-repair are independent
            exponential random variables
        with the same mean.

    */

    double ct1, ct, al;
    long k, n, m, nta, ntd;
    long i, j;   /* work variables */
    double p,w;
    double unirv(void);
    /* initialize variables */
    k = 1000000;
    ct1 = 0.0;
    ct = 0.0;
    n = 1;   /* assume one machine in repair at
        the start of simulation */
    m = 3;
    nta = 0;
    ntd = 0;
```

```
/* main simulation loop */
for( i =1;  i <=k;  i ++) {
   if  (m ==  4) {
      /*failure when all machines are
         currently working*/
      ct = ct1 − (2.0/m)*log(unirv());
      a1=a1+n*(ct−ct1);
      n=n+1;
      m=m−1;
      nta=nta+1;

   } else {
      /*determine which kind of event failure
         or repair completion */
      ct=ct1 −(2.0/(m+1.0))*log(unirv());
      p = (double) m/(m+1);
      if (unirv()<p) {
         /*this is a machine failure */
         a1=a1+n*(ct−ct1);   /*accumulate time
            weighed system size */
         n=n+1;                 /*up number of
            machines in repair */
         m=m−1;                 /*decrease number
            of machines operating */
         nta=nta+1;             /*record arrival to
            repair facility */
      } else {
         /*this is a repair completion*/
         a1=a1+n*(ct−ct1);   /* accumulate time
            weighted system size */
         ntd=ntd+1;             /*record repair
            completion */
         n=n−1;                 /*decrease number
            at repair facility */
         m=m+1;                 /*increase number
            of machines in operation */
      }
   }
   ct1=ct;
} /* end of for−loop */

/*calculate simulation results*/
a1 = a1/ct1;        /*average number of machines
   in repair*/
```

```
w = ctl*al/nta;   /*use Little's Law to compute
    average time in repair*/

/*output results*/
printf("Mean number of machines in repair = %
    lf vs theoretical value of 3.01538\n",al);
printf("Mean repair time = %lf vs theoretical
    value of 6.125\n",w);

return 0;
}

double unirv(void) {
/*function to return uniform random variable
    [0,1)*/
return rand()/((double)RAND_MAX+1);
}
```

As sample output, the program estimates that the mean number of machines in repair is 3.015975 versus the theoretical value of 3.01538. The estimated mean repair time is 6.121536 versus the theoretical value of 6.125.

The next program is written in Visual Basic for Applications (VBA) in Excel. This version explicitly models the failure times and repair times for individual machines. With this explicit modeling, this version can easily be extended to the situation where the means for the time to failure and the time to repair have different values.

```
Option Explicit
Const INFINITY = 1.79769313486231E+308 'Largest
    number allowed... essentially infinity for this
        program
Const MTBF = 2#     'Mean-time-between-failurers
Const MTTR = 2#     'Mean-time-to-repair
Const NUMBER_OF_MACHINES = 4
Const NUMBER_OF_RUNS = 1000000     'run length
    counter

'The following variables are used:
'   dEventTime() vector of event times
'       dEventTime(0) is for repair completion,
'       dEventTime(1 to NUMBER_OF_MACHINES) failure
    time for machines 1 to NUMBER_OF_MACHINES
'   lRun counter of the number of simulation runs
'   dCurrentClock is the current time of the
    simulated system
```

' dFailureDetectedTime is the time for the next
  failure
' dRepairCompletedTime is the time a machine
  completes repair at the service facility
' dSystemSizeAccumulator is an accumulator for
  mean system size
' dMeanTimeForRepair is mean to repair a down
  machine
' lOperatingMachines is the number of machines
  in operation
' lDownMachines is the number of machines down,
  in need of repair
'   Note: NumberOfOperatingMachines +
  NumberOfDownMachines = 4
' lNumberOfServiceBreakdowns number of arrivals
  to the service facility
' lNumberOfRepairs number of departures or
  repairs from the service facility

```
Sub MachineRepairModel()
    Dim dEventTime(0 To NUMBER_OF_MACHINES) As
        Double
    Dim dSystemSizeAccumulator As Double,
        dMeanTimeForRepair As Double
    Dim lOperatingMachines As Long, lDownMachines
        As Long
    Dim lNumberOfServiceBreakdowns As Long,
        lNumberOfRepairs As Long
    Dim dCurrentClock As Double
    Dim lRun As Long       'run length counter
    Dim iNextEventIndex As Integer, i As Integer
    Dim dTotalDelta As Double

    'Initialize variables
    dSystemSizeAccumulator = 0
    lOperatingMachines = 4
    lDownMachines = 0
    lNumberOfServiceBreakdowns = 0
    lNumberOfRepairs = 0
    dCurrentClock = 0
    dEventTime(0) = INFINITY      'no repairs at
        start of simulation
    For i = 1 To NUMBER_OF_MACHINES
        dEventTime(i) = -MTBF * Log(Rnd())    '
            failure time for Machine i
```

```
Next i

'Perform main simulation loop
For lRun = 1 To NUMBER_OF_RUNS

    'determine next event to process, i.e.,
        find the minimum event time
    'assume next event is EventTime(0) entry
    iNextEventIndex = 0
    For i = 1 To UBound(dEventTime)
        If dEventTime(i) < dEventTime(
            iNextEventIndex) Then
            'earlier time found, reset to
                earlier time
            iNextEventIndex = i
        End If
    Next i

    'determine if next event is a machine
        failure or repair completion
    If iNextEventIndex > 0 Then
        'next event is failure
        'update counters
        dSystemSizeAccumulator =
            dSystemSizeAccumulator + CDbl(
            lDownMachines) * (dEventTime(
            iNextEventIndex) - dCurrentClock)
        lNumberOfServiceBreakdowns =
            lNumberOfServiceBreakdowns + 1
        lOperatingMachines =
            lOperatingMachines - 1
        lDownMachines = lDownMachines + 1

        'update clock to current event
        dCurrentClock = dEventTime(
            iNextEventIndex)

        'signal machine in repair
        dEventTime(iNextEventIndex) =
            INFINITY

        If lDownMachines = 1 Then 'schedule
            repair for machine entering empty
            repair facility
```

```
            dEventTime(0) = dCurrentClock - (
                MTTR * Log(Rnd()))
        End If

    Else
        'next event is repair completion
        'update counters
        dSystemSizeAccumulator =
            dSystemSizeAccumulator + CDbl(
            lDownMachines) * (dEventTime(0) -
            dCurrentClock)
        lNumberOfRepairs = lNumberOfRepairs +
            1
        lDownMachines = lDownMachines - 1
        lOperatingMachines =
            lOperatingMachines + 1

        'update clock to current time
        dCurrentClock = dEventTime(0)

        'determine time for repair completion
        If lDownMachines > 0 Then
            dEventTime(0) = dCurrentClock - (
                MTTR * Log(Rnd()))
        Else
            dEventTime(0) = INFINITY
        End If

        'schedule next failure for machine
            fixed
        For i = 1 To NUMBER_OF_MACHINES
            If dEventTime(i) = INFINITY Then
                dEventTime(i) = dCurrentClock
                    - (MTBF * Log(Rnd()))
                Exit For
            End If
        Next i

    End If
Next lRun

'compute simulation statistics
dSystemSizeAccumulator =
    dSystemSizeAccumulator / dCurrentClock
```

```
dMeanTimeForRepair = dCurrentClock *
    dSystemSizeAccumulator /
    lNumberOfServiceBreakdowns

'print results in a pop-up box
MsgBox "Mean number of machines in repair ="
    & dSystemSizeAccumulator & " vs.
        theoretical value of 3.01538" & vbCrLf _
            & "Mean repair time = " &
                dMeanTimeForRepair & " vs.
                theoretical value of 6.125"
```

End Sub

Sample output for the VBA program follows.

8.15   Let $d_i = \overline{W}_{i1} - \overline{W}_{i2}$ be the paired difference of the mean waiting times for replication i. The sample mean of the differences is

$$\bar{d} = \sum_{i=1}^{15} d_i/15 \doteq -0.7147.$$

The sample standard deviation of the differences is

$$s_{\bar{d}} \doteq 1.2076.$$

Using the t-distribution, a 95% confidence interval on the true difference of the two designs is

$$\bar{d} \pm t_{14}(0.025)s_{\bar{d}}/\sqrt{15} = -0.7147 \pm 2.1448 \cdot 1.2076/\sqrt{15}$$
$$= (-1.383, -0.046).$$

Since the interval does not include 0, we reject the hypothesis of no difference. Design 1 may be preferable.

8.18   Using direct integration,

$$f^*(s) = \int_0^\infty e^{-st} f(t)\, dt = 4\mu^2 \int_0^\infty te^{-(s+2\mu)t}\, dt.$$

Using integration by parts,

$$f^*(s) = 4\mu^2 \left[\frac{te^{-(s+2\mu)t}}{-(s+2\mu)} \bigg|_{t=0}^{t=\infty} + \frac{1}{s+2\mu} \int_0^\infty e^{-(s+2\mu)t} \, dt \right]$$

$$= 4\mu^2 \left[0 + \frac{1}{(s+2\mu)^2} \right]$$

$$= \frac{2\mu}{s+2\mu} \cdot \frac{2\mu}{s+2\mu}.$$

This is the same as the product of the LST of two exponential random variables, each with mean $1/2\mu$.

8.21   **(a)** We show the result in reverse:

$$\mathrm{Re} \left[\frac{e^{bt}}{\pi} \int_{-\infty}^\infty \frac{\lambda(b+\lambda)}{(b+\lambda)^2 + z^2} e^{izt} \, dz \right]$$

$$= \mathrm{Re} \left[\frac{e^{bt}}{\pi} \int_{-\infty}^\infty \frac{\lambda(b+\lambda)}{(b+\lambda)^2 + z^2} [\cos(zt) + i\sin(zt)] \, dz \right]$$

$$= \frac{e^{bt}}{\pi} \int_{-\infty}^\infty \frac{\lambda(b+\lambda)}{(b+\lambda)^2 + z^2} \cos(zt) \, dz$$

$$= \frac{2e^{bt}}{\pi} \int_0^\infty \frac{\lambda(b+\lambda)}{(b+\lambda)^2 + z^2} \cos(zt) \, dz.$$

The last equality follows since the integrand is symmetric about 0.

(b) The integral along the base of the semicircular region is

$$f(t) = \mathrm{Re} \left[\frac{e^{bt}}{\pi} \int_{-R}^R \frac{\lambda(b+\lambda)}{(b+\lambda)^2 + z^2} e^{izt} \, dz \right],$$

which is close to the desired result for large R. Thus it remains to show that the integral on the semicircular part can be made arbitrarily small. On this part, the curve can be parameterized by $z = Re^{i\theta}$ for $\theta \in [0, \pi]$. For such z, the integrand is of order $1/R^2$. The length of the path is πR, so the integral is of order $1/R$. Thus for large R, the integral along this part of the path can be made arbitrarily small.

(c) The integrand has poles at $z = \pm(b+\lambda)i$. Only one of the poles, namely $z = (b+\lambda)i$, lies within the semicircular contour. Its residue is

$$(z - (b+\lambda)i)\frac{\lambda(b+\lambda)}{(b+\lambda)^2 + z^2} e^{izt} \bigg|_{z=(b+\lambda)i} = \frac{\lambda}{2i} e^{-(b+\lambda)t}.$$

(d) Using the residue theorem from complex variables,

$$f(t) = \mathrm{Re} \left[\frac{e^{bt}}{\pi} \cdot 2\pi i \cdot \frac{\lambda}{2i} e^{-(b+\lambda)t} \right] = \lambda e^{-\lambda t}.$$

The result does not depend on the value of b.

WILEY SERIES IN PROBABILITY AND STATISTICS
ESTABLISHED BY WALTER A. SHEWHART AND SAMUEL S. WILKS

Editors: *David J. Balding, Noel A. C. Cressie, Garrett M. Fitzmaurice,*
Iain M. Johnstone, Geert Molenberghs, David W. Scott, Adrian F. M. Smith,
Ruey S. Tsay, Sanford Weisberg
Editors Emeriti: *Vic Barnett, J. Stuart Hunter, Jozef L. Teugels*

The *Wiley Series in Probability and Statistics* is well established and authoritative. It covers many topics of current research interest in both pure and applied statistics and probability theory. Written by leading statisticians and institutions, the titles span both state-of-the-art developments in the field and classical methods.

Reflecting the wide range of current research in statistics, the series encompasses applied, methodological and theoretical statistics, ranging from applications and new techniques made possible by advances in computerized practice to rigorous treatment of theoretical approaches.

This series provides essential and invaluable reading for all statisticians, whether in academia, industry, government, or research.

*Now available in a lower priced paperback edition in the Wiley Classics Library.
†Now available in a lower priced paperback edition in the Wiley–Interscience Paperback Series.

*Now available in a lower priced paperback edition in the Wiley Classics Library.
†Now available in a lower priced paperback edition in the Wiley–Interscience Paperback Series.

* COCHRAN and COX · Experimental Designs, *Second Edition*
CONGDON · Applied Bayesian Modelling
CONGDON · Bayesian Models for Categorical Data
CONGDON · Bayesian Statistical Modelling
CONOVER · Practical Nonparametric Statistics, *Third Edition*
COOK · Regression Graphics
COOK and WEISBERG · Applied Regression Including Computing and Graphics
COOK and WEISBERG · An Introduction to Regression Graphics
CORNELL · Experiments with Mixtures, Designs, Models, and the Analysis of Mixture
    Data, *Third Edition*
COVER and THOMAS · Elements of Information Theory
COX · A Handbook of Introductory Statistical Methods
* COX · Planning of Experiments
CRESSIE · Statistics for Spatial Data, *Revised Edition*
CSÖRGŐ and HORVÁTH · Limit Theorems in Change Point Analysis
DANIEL · Applications of Statistics to Industrial Experimentation
DANIEL · Biostatistics: A Foundation for Analysis in the Health Sciences, *Eighth Edition*
* DANIEL · Fitting Equations to Data: Computer Analysis of Multifactor Data,
    Second Edition
DASU and JOHNSON · Exploratory Data Mining and Data Cleaning
DAVID and NAGARAJA · Order Statistics, *Third Edition*
* DEGROOT, FIENBERG, and KADANE · Statistics and the Law
DEL CASTILLO · Statistical Process Adjustment for Quality Control
DeMARIS · Regression with Social Data: Modeling Continuous and Limited Response
    Variables
DEMIDENKO · Mixed Models: Theory and Applications
DENISON, HOLMES, MALLICK and SMITH · Bayesian Methods for Nonlinear
    Classification and Regression
DETTE and STUDDEN · The Theory of Canonical Moments with Applications in
    Statistics, Probability, and Analysis
DEY and MUKERJEE · Fractional Factorial Plans
DILLON and GOLDSTEIN · Multivariate Analysis: Methods and Applications
DODGE · Alternative Methods of Regression
* DODGE and ROMIG · Sampling Inspection Tables, *Second Edition*
* DOOB · Stochastic Processes
DOWDY, WEARDEN, and CHILKO · Statistics for Research, *Third Edition*
DRAPER and SMITH · Applied Regression Analysis, *Third Edition*
DRYDEN and MARDIA · Statistical Shape Analysis
DUDEWICZ and MISHRA · Modern Mathematical Statistics
DUNN and CLARK · Basic Statistics: A Primer for the Biomedical Sciences,
    Third Edition
DUPUIS and ELLIS · A Weak Convergence Approach to the Theory of Large Deviations
EDLER and KITSOS · Recent Advances in Quantitative Methods in Cancer and Human
    Health Risk Assessment
* ELANDT-JOHNSON and JOHNSON · Survival Models and Data Analysis
  ENDERS · Applied Econometric Time Series
† ETHIER and KURTZ · Markov Processes: Characterization and Convergence
EVANS, HASTINGS, and PEACOCK · Statistical Distributions, *Third Edition*
FELLER · An Introduction to Probability Theory and Its Applications, Volume I,
    Third Edition, Revised; Volume II, *Second Edition*
FISHER and VAN BELLE · Biostatistics: A Methodology for the Health Sciences
FITZMAURICE, LAIRD, and WARE · Applied Longitudinal Analysis
* FLEISS · The Design and Analysis of Clinical Experiments

*Now available in a lower priced paperback edition in the Wiley Classics Library.
†Now available in a lower priced paperback edition in the Wiley–Interscience Paperback Series.

FLEISS · Statistical Methods for Rates and Proportions, *Third Edition*
† FLEMING and HARRINGTON · Counting Processes and Survival Analysis
FULLER · Introduction to Statistical Time Series, *Second Edition*
† FULLER · Measurement Error Models
GALLANT · Nonlinear Statistical Models
GEISSER · Modes of Parametric Statistical Inference
GELMAN and MENG · Applied Bayesian Modeling and Causal Inference from Incomplete-Data Perspectives
GEWEKE · Contemporary Bayesian Econometrics and Statistics
GHOSH, MUKHOPADHYAY, and SEN · Sequential Estimation
GIESBRECHT and GUMPERTZ · Planning, Construction, and Statistical Analysis of Comparative Experiments
GIFI · Nonlinear Multivariate Analysis
GIVENS and HOETING · Computational Statistics
GLASSERMAN and YAO · Monotone Structure in Discrete-Event Systems
GNANADESIKAN · Methods for Statistical Data Analysis of Multivariate Observations, *Second Edition*
GOLDSTEIN and LEWIS · Assessment: Problems, Development, and Statistical Issues
GREENWOOD and NIKULIN · A Guide to Chi-Squared Testing
GROSS, SHORTLE, THOMPSON, and HARRIS · Fundamentals of Queueing Theory, *Fourth Edition*
GROSS, SHORTLE, THOMPSON, and HARRIS · Solutions Manual to Accompany Fundamentals of Queueing Theory, *Fourth Edition*
* HAHN and SHAPIRO · Statistical Models in Engineering
HAHN and MEEKER · Statistical Intervals: A Guide for Practitioners
HALD · A History of Probability and Statistics and their Applications Before 1750
HALD · A History of Mathematical Statistics from 1750 to 1930
† HAMPEL · Robust Statistics: The Approach Based on Influence Functions
HANNAN and DEISTLER · The Statistical Theory of Linear Systems
HARTUNG, KNAPP, and SINHA · Statistical Meta-Analysis with Applications
HEIBERGER · Computation for the Analysis of Designed Experiments
HEDAYAT and SINHA · Design and Inference in Finite Population Sampling
HEDEKER and GIBBONS · Longitudinal Data Analysis
HELLER · MACSYMA for Statisticians
HINKELMANN and KEMPTHORNE · Design and Analysis of Experiments, Volume 1: Introduction to Experimental Design, *Second Edition*
HINKELMANN and KEMPTHORNE · Design and Analysis of Experiments, Volume 2: Advanced Experimental Design
HOAGLIN, MOSTELLER, and TUKEY · Exploratory Approach to Analysis of Variance
* HOAGLIN, MOSTELLER, and TUKEY · Exploring Data Tables, Trends and Shapes
* HOAGLIN, MOSTELLER, and TUKEY · Understanding Robust and Exploratory Data Analysis
HOCHBERG and TAMHANE · Multiple Comparison Procedures
HOCKING · Methods and Applications of Linear Models: Regression and the Analysis of Variance, *Second Edition*
HOEL · Introduction to Mathematical Statistics, *Fifth Edition*
HOGG and KLUGMAN · Loss Distributions
HOLLANDER and WOLFE · Nonparametric Statistical Methods, *Second Edition*
HOSMER and LEMESHOW · Applied Logistic Regression, *Second Edition*
HOSMER, LEMESHOW, and MAY · Applied Survival Analysis: Regression Modeling of Time-to-Event Data, *Second Edition*
† HUBER · Robust Statistics

*Now available in a lower priced paperback edition in the Wiley Classics Library.
†Now available in a lower priced paperback edition in the Wiley–Interscience Paperback Series.

*Now available in a lower priced paperback edition in the Wiley Classics Library.

†Now available in a lower priced paperback edition in the Wiley–Interscience Paperback Series.

---

*Now available in a lower priced paperback edition in the Wiley Classics Library.

†Now available in a lower priced paperback edition in the Wiley–Interscience Paperback Series.

*Now available in a lower priced paperback edition in the Wiley Classics Library.
†Now available in a lower priced paperback edition in the Wiley–Interscience Paperback Series.

*Now available in a lower priced paperback edition in the Wiley Classics Library.
†Now available in a lower priced paperback edition in the Wiley–Interscience Paperback Series.

VESTRUP · The Theory of Measures and Integration

VIDAKOVIC · Statistical Modeling by Wavelets

VINOD and REAGLE · Preparing for the Worst: Incorporating Downside Risk in Stock Market Investments

WALLER and GOTWAY · Applied Spatial Statistics for Public Health Data

WEERAHANDI · Generalized Inference in Repeated Measures: Exact Methods in MANOVA and Mixed Models

WEISBERG · Applied Linear Regression, *Third Edition*

WELSH · Aspects of Statistical Inference

WESTFALL and YOUNG · Resampling-Based Multiple Testing: Examples and Methods for *p*-Value Adjustment

WHITTAKER · Graphical Models in Applied Multivariate Statistics

WINKER · Optimization Heuristics in Economics: Applications of Threshold Accepting

WONNACOTT and WONNACOTT · Econometrics, *Second Edition*

WOODING · Planning Pharmaceutical Clinical Trials: Basic Statistical Principles

WOODWORTH · Biostatistics: A Bayesian Introduction

WOOLSON and CLARKE · Statistical Methods for the Analysis of Biomedical Data, *Second Edition*

WU and HAMADA · Experiments: Planning, Analysis, and Parameter Design Optimization

WU and ZHANG · Nonparametric Regression Methods for Longitudinal Data Analysis

YANG · The Construction Theory of Denumerable Markov Processes

YOUNG, VALERO-MORA, and FRIENDLY · Visual Statistics: Seeing Data with Dynamic Interactive Graphics

ZELTERMAN · Discrete Distributions—Applications in the Health Sciences

* ZELLNER · An Introduction to Bayesian Inference in Econometrics

ZHOU, OBUCHOWSKI, and McCLISH · Statistical Methods in Diagnostic Medicine

*Now available in a lower priced paperback edition in the Wiley Classics Library.

†Now available in a lower priced paperback edition in the Wiley–Interscience Paperback Series.

Printed and bound by CPI Group (UK) Ltd, Croydon, CR0 4YY

27/10/2024

14580263-0002